Atmospheric Science:
Understanding the Earth's Atmosphere

Atmospheric Science: Understanding the Earth's Atmosphere

Eleanor Flores

SYRAWOOD
PUBLISHING HOUSE
New York

Published by Syrawood Publishing House,
750 Third Avenue, 9th Floor,
New York, NY 10017, USA
www.syrawoodpublishinghouse.com

Atmospheric Science: Understanding the Earth's Atmosphere
Eleanor Flores

International Standard Book Number: 978-1-64740-107-8 (Hardback)

Cataloging-in-Publication Data

Atmospheric science : understanding the earth's atmosphere / Eleanor Flores.
p. cm.
Includes bibliographical references and index.
ISBN 978-1-64740-107-8
1. Atmospheric science. 2. Atmosphere. 3. Earth sciences. 4. Meteorology.
5. Climatology. I. Flores, Eleanor.
QC866 .A86 2022
551.5--dc23

Table of Contents

Preface

It is with great pleasure that I present this book. It has been carefully written after numerous discussions with my peers and other practitioners of the field. I would like to take this opportunity to thank my family and friends who have been extremely supporting at every step in my life.

The study of the Earth's atmosphere and the related physical processes falls under the domain of atmospheric science. It can be divided into meteorology, aeronomy and climatology. The discipline of meteorology includes atmospheric chemistry and atmospheric physics, and primarily focuses on weather forecasting. Climatology studies the changes and variations occurring in the atmosphere. Aeronomy studies the upper layers of the atmosphere with a major focus upon dissociation and ionization. Atmospheric science also includes planetary science, and the study of the atmosphere of planets and natural satellites of the solar system. This textbook is compiled in such a manner, that it will provide in-depth knowledge about the theory and practice of this field. The various sub-fields of atmospheric science along with technological progress that have future implications are glanced at in this book. It is a complete source of knowledge on the present status of this important discipline.

The chapters below are organized to facilitate a comprehensive understanding of the subject:

Chapter - Introduction

Atmospheric science is a domain that deals with the study of the atmosphere of earth as well as its diverse inner working physical processes. It makes use of various tools such as differential equations and vector analysis. This is an introductory chapter which will introduce briefly all the significant aspects of atmospheric science.

Chapter - Branches of Atmospheric Science

Atmospheric science is a vast discipline that branches into several sub-disciplines. Some of these are climatology, meteorology, atmospheric chemistry, atmospheric dynamics, coniology and atmospheric physics. This chapter has been carefully written to provide an easy understanding of these branches of atmospheric science.

Chapter - Layers of Earth's Atmosphere

The atmosphere of the Earth is divided into various layers, all of which have their own specific traits. Some of these layers are the troposphere, stratosphere, mesosphere, thermosphere, exosphere, ozonosphere and magnetosphere. This chapter has been carefully written to provide an easy understanding of these layers of the Earth's atmosphere.

Chapter - Atmospheric Phenomenon

Atmospheric phenomena are the observable events that are the outcome of the interactions of light and matter. It includes phenomena like belt of venus, green flash, 22 degree halos, thunderstorm, jet stream, etc. All the diverse principles of these atmospheric phenomena have been carefully analyzed in this chapter.

Chapter - Key Concepts in Atmospheric Science

Atmospheric temperature, atmospheric refraction, atmospheric density, atmospheric circulation, atmospheric electricity, diffuse sky radiation and atmospheric pressure are some of the key concepts of atmospheric science. The chapter closely examines these key concepts of atmospheric science to provide an extensive understanding of the subject.

Chapter - Weather Forecasting

The application of science and technology to predict the atmospheric conditions of a given location and time is referred to as weather forecasting. Nowcasting, marine weather forecasting, numerical weather prediction, etc. are some of the various types of weather forecasting. This chapter discusses in detail these types of weather forecasting.

Chapter - Climate Change and Global Warming

Climate change is a condition when the changes in Earth's climate system cause new weather patterns. The long term increase in the average temperature of the Earth's climate system is termed as global warming. The chapter closely examines the key concepts related to climate change such as ozone depletion and global warming to provide an extensive understanding of the subject.

Eleanor Flores

1

Introduction

Atmospheric science is a domain that deals with the study of the atmosphere of earth as well as its diverse inner working physical processes. It makes use of various tools such as differential equations and vector analysis. This is an introductory chapter which will introduce briefly all the significant aspects of atmospheric science.

ATMOSPHERE OF EARTH

Earth's atmosphere is a thin blanket of gases and tiny particles — together called air. We are most aware of air when it moves and creates wind. All living things need some of the gases in air for life support. Without an atmosphere, Earth would likely be just another lifeless rock.

Earth's atmosphere, along with the abundant liquid water at Earth's surface, are the keys to our planet's unique place in the solar system. Much of what makes Earth exceptional depends on the atmosphere.

Indispensable for Life on Earth

Without the atmosphere, Earth would look a lot more like the Moon. Atmospheric gases, especially carbon dioxide (CO_2) and oxygen (O_2), are extremely important for living organisms. How does the atmosphere make life possible? How does life alter the atmosphere?

In photosynthesis plants use CO_2 and create O_2. Photosynthesis is responsible for nearly all of the oxygen currently found in the atmosphere. By creating oxygen and food, plants have made an environment that is favorable for animals. In respiration, animals use oxygen to convert sugar into food energy they can use. Plants also go through respiration and consume some of the sugars they produce.

Crucial Part of the Water Cycle

As part of the hydrologic cycle, which was detailed in the Earth's Fresh Water chapter, water spends a lot of time in the atmosphere, mostly as water vapor.All weather takes place in the atmosphere, virtually all of it in the lower atmosphere. Weather describes what the atmosphere is like at a specific time and place, and may include temperature, wind, and precipitation. Weather is the change we experience from day to day. Climate is the long-term average of weather in a particular spot. Although the weather for a particular winter day in Tucson, Arizona, may include snow, the climate of Tucson is generally warm and dry.

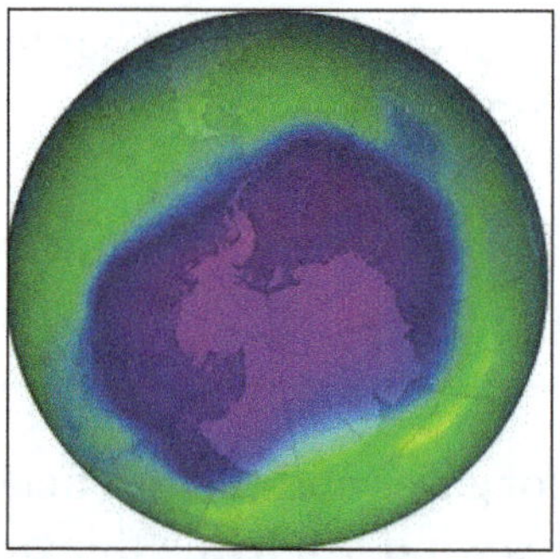

Ozone Layer Makes Life Possible

Ozone is a molecule composed of three oxygen atoms, (O_3). Ozone in the upper atmosphere absorbs high-energy ultraviolet (UV) radiation coming from the Sun. This protects living things on Earth's surface from the Sun's most harmful rays. Without ozone for protection, only the simplest life forms would be able to live on Earth.

Moderates Earth's Temperature

Along with the oceans, the atmosphere keeps Earth's temperatures within an acceptable range.Greenhouse gases trap heat in the atmosphere so they help to moderate global temperatures. Without an atmosphere with greenhouse gases, Earth's temperatures would be frigid at night and scorching during the day. Important greenhouse gases include carbon dioxide, methane, water vapor, and ozone.

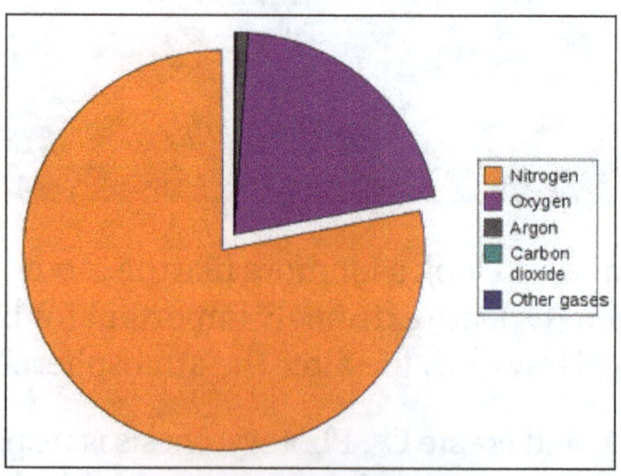

Atmospheric Gases.

Composition of the Atmosphere

Nitrogen and oxygen together make up 99 percent of the planet's atmosphere. The rest of the gases are minor components but sometimes are very important. Humidity is the amount of water vapor in the air. Humidity varies from place to place and season to season. This fact is obvious if you compare a summer day in Atlanta, Georgia, where humidity is high, with a winter day in Phoenix, Arizona, where humidity is low. When the air is very humid, it feels heavy or sticky. Dry air usually feels more comfortable.Where around the globe is mean atmospheric water vapor higher and where is it lower and why? Higher humidity is found around the equatorial regions because air temperatures are higher and warm air can hold more moisture than cooler air. Of course, humidity is lower near the polar regions because air temperature is lower.

Some of what is in the atmosphere is not gas. Particles of dust, soil, fecal matter, metals, salt, smoke, ash, and other solids make up a small percentage of the atmosphere. Particles provide starting points (or nuclei) for water vapor to condense on and form raindrops.

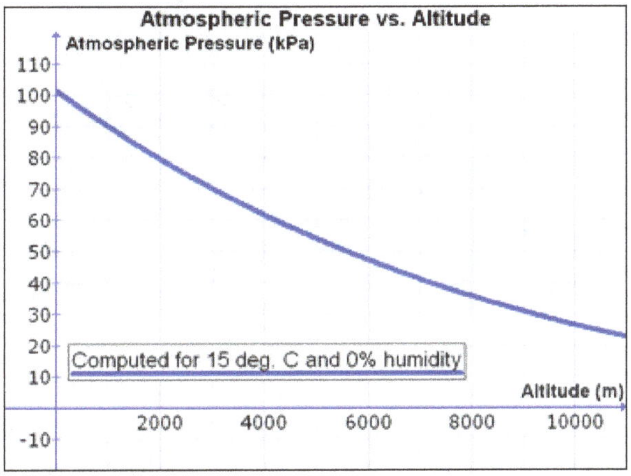

Atmospheric Pressure and Density

The atmosphere has different properties at different elevations above sea level, or altitudes. The air density (the number of molecules in a given volume) decreases with increasing altitude. This is why people who climb tall mountains, such as Mt. Everest, have to set up camp at different elevations to let their bodies get used to the decreased air.Why does air density decrease with altitude? Gravity pulls the gas molecules towards Earth's center. The pull of gravity is stronger closer to the center at sea level. Air is denser at sea level where the gravitational pull is greater.Gases at sea level are also compressed by the weight of the atmosphere above them. The force of the air weighing down over a unit of area is known as its atmospheric pressure. The reason why we are not crushed by this weight is because the molecules inside our bodies are pushing outward to compensate. Atmospheric pressure is felt from all directions, not just from above.

At higher altitudes the atmospheric pressure is lower and the air is less dense than at higher altitudes. If your ears have ever "popped", you have experienced a change in air pressure. Gas molecules are found inside and outside your ears. When you change altitude quickly, like when an airplane is descending, your inner ear keeps the density of molecules at the original altitude. Eventually the air molecules inside your ear suddenly move through a small tube in your ear to equalize the pressure. This sudden rush of air is felt as a popping sensation.

Although the density of the atmosphere changes with altitude, the composition stays the same with altitude, with one exception. In the ozone layer, at about 20 km to 40 km above the surface, there is a greater concentration of ozone molecules than in other portions of the atmosphere.

Layers of the Atmosphere

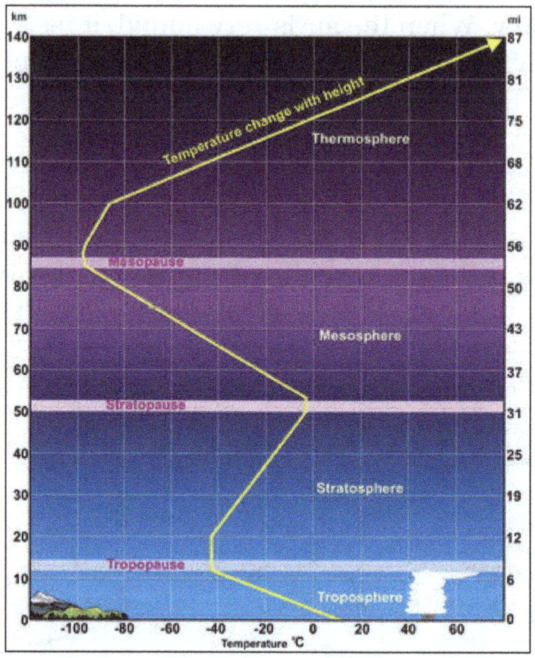

The atmosphere is layered, corresponding with how the atmosphere's temperature changes with altitude. By understanding the way temperature changes with altitude, we can learn a lot about how the atmosphere works. While weather takes place in the lower atmosphere, interesting things, such as the beautiful aurora, happen higher in the atmosphere.

Why does warm air rise? Gas molecules are able to move freely and if they are uncontained, as they are in the atmosphere, they can take up more or less space.

- When gas molecules are cool, they are sluggish and do not take up as much space. With the same number of molecules in less space, both air density and air pressure are higher.
- When gas molecules are warm, they move vigorously and take up more space. Air density and air pressure are lower.

Warmer, lighter air is more buoyant than the cooler air above it, so it rises. The cooler air then sinks down, because it is denser than the air beneath it.

The property that changes most strikingly with altitude is air temperature. Unlike the change in pressure and density, which decrease with altitude, changes in air temperature are not regular. A change in temperature with distance is called a temperature gradient.

The atmosphere is divided into layers based on how the temperature in that layer changes with altitude, the layer's temperature gradient. The temperature gradient of each layer is different. In some layers, temperature increases with altitude and in others it decreases. The temperature gradient in each layer is determined by the heat source of the layer. Most of the important processes of the atmosphere take place in the lowest two layers: the troposphere and the stratosphere.

Troposphere

The temperature of the troposphere is highest near the surface of the Earth and decreases with altitude. On average, the temperature gradient of the troposphere is 6.5oC per 1,000 m (3.6oF per 1,000 ft.) of altitude. What is the source of heat for the troposphere? Earth's surface is a major source of heat for the troposphere, although nearly all of that heat comes from the Sun. Rock, soil, and water on Earth absorb the Sun's light and radiate it back into the atmosphere as heat. The temperature is also higher near the surface because of the greater density of gases.

Notice that in the troposphere warmer air is beneath cooler air. What do you think the consequence of this is? This condition is unstable. The warm air near the surface rises and cool air higher in the troposphere sinks. So air in the troposphere does a lot of mixing. This mixing causes the temperature gradient to vary with time and place. The rising and sinking of air in the troposphere means that all of the planet's weather takes place in the troposphere.

Sometimes there is a temperature inversion, air temperature in the troposphere increases with altitude and warm air sits over cold air. Inversions are very stable and may last for several days or even weeks. They form:

- Over land at night or in winter when the ground is cold. The cold ground cools the air that sits above it, making this low layer of air denser than the air above it.
- Near the coast where cold seawater cools the air above it. When that denser air moves inland, it slides beneath the warmer air over the land.

Since temperature inversions are stable, they often trap pollutants and produce unhealthy air conditions in cities. At the top of the troposphere is a thin layer in which the temperature does not change with height. This means that the cooler, denser air of the troposphere is trapped beneath the warmer, less dense air of the stratosphere. Air from the troposphere and stratosphere rarely mix.

Stratosphere

Ash and gas from a large volcanic eruption may burst into the stratosphere, the layer above the troposphere. Once in the stratosphere, it remains suspended there for many years because there is so little mixing between the two layers. Pilots like to fly in the lower portions of the stratosphere because there is little air turbulence.In the stratosphere, temperature increases with altitude. What is the heat source for the stratosphere? The direct heat source for the stratosphere is the Sun. Air in the stratosphere is stable because warmer, less dense air sits over cooler, denser air. As a result, there is little mixing of air within the layer.The ozone layer is found within the stratosphere between 15 to 30 km (9 to 19 miles) altitude. The thickness of the ozone layer varies by the season and also by latitude.

The ozone layer is extremely important because ozone gas in the stratosphere absorbs most of the Sun's harmful ultraviolet (UV) radiation. Because of this, the ozone layer protects life on Earth. High-energy UV light penetrates cells and damages DNA, leading to cell death (which we know as a bad sunburn). Organisms on Earth are not adapted to heavy UV exposure, which kills or damages them. Without the ozone layer to reflect UVC and UVB radiation, most complex life on Earth would not survive long.

Mesosphere

Temperatures in the mesosphere decrease with altitude. Because there are few gas molecules in the mesosphere to absorb the Sun's radiation, the heat source is the stratosphere below. The mesosphere is extremely cold, especially at its top, about -90 degrees C (-130 degrees F).

The air in the mesosphere has extremely low density: 99.9 percent of the mass of the atmosphere is below the mesosphere. As a result, air pressure is very low. A person traveling through the mesosphere would experience severe burns from ultraviolet light since the ozone layer which provides UV protection is in the stratosphere below. There would be almost no oxygen for breathing. Stranger yet, an unprotected traveler's blood would boil at normal body temperature because the pressure is so low.

Thermosphere

The density of molecules is so low in the thermosphere that one gas molecule can go about 1 km before it collides with another molecule. Since so little energy is transferred, the air feels very cold. Within the thermosphere is the ionosphere. The ionosphere gets its name from the solar radiation that ionizes gas molecules to create a positively charged ion and one or more negatively charged electrons. The freed electrons travel within the ionosphere as electric currents. Because of the free ions, the ionosphere has many interesting characteristics. At night, radio waves bounce off the ionosphere and back to Earth. This is why you can often pick up an AM radio station far from its source at night.The Van Allen radiation belts are two doughnut-shaped zones of highly charged particles that are located beyond the atmosphere in the magnetosphere. The particles originate in solar flares and fly to Earth on the solar wind. Once trapped by Earth's magnetic field, they follow along the field's magnetic lines of force. These lines extend from above the equator to the North Pole and also to the South Pole then return to the equator.

When massive solar storms cause the Van Allen belts to become overloaded with particles, the result is the most spectacular feature of the ionosphere — the aurora. The particles spiral along magnetic field lines toward the poles. The charged particles energize oxygen and nitrogen gas molecules, causing them to light up. Each gas emits a particular color of light.

There is no real outer limit to the exosphere, the outermost layer of the atmosphere; the gas molecules finally become so scarce that at some point there are no more. Beyond the atmosphere is the solar wind. The solar wind is made of high-speed particles, mostly protons and electrons, traveling rapidly outward from the Sun.

Atmospheric Energy, Temperature and Heat

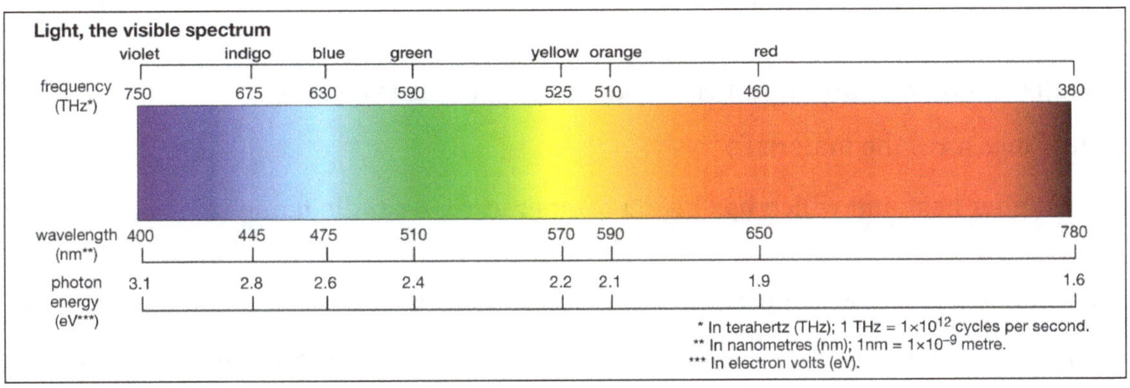

Energy

Energy travels through space or material. This is obvious when you stand near a fire and feel its warmth or when you pick up the handle of a metal pot even though the handle is not sitting directly on the hot stove. Invisible energy waves can travel through air, glass, and even the vacuum of outer space. These waves have electrical and magnetic properties, so they are called electromagnetic waves. The transfer of energy from one object to another through electromagnetic waves is known as radiation. Different wavelengths of energy create different types of electromagnetic waves.

- The wavelengths humans can see are known as "visible light." These wavelengths appear to us as the colors of the rainbow. What objects can you think of that radiate visible light? Two include the Sun and a light bulb.
- The longest wavelengths of visible light appear red. Infrared wavelengths are longer than visible red. Snakes can see infrared energy. We feel infrared energy as heat.
- Wavelengths that are shorter than violet are called ultraviolet.

Can you think of some objects that appear to radiate visible light, but actually do not? The moon and the planets do not emit light of their own; they reflect the light of the Sun. Reflection is when light (or another wave) bounces back from a surface. Albedo is a measure of how well a surface reflects light. A surface with high albedo reflects a large percentage of light. A snow field has high albedo.

One important fact to remember is that energy cannot be created or destroyed — it can only be changed from one form to another. This is such a fundamental fact of nature that it is a law: the law of conservation of energy.

In photosynthesis, for example, plants convert solar energy into chemical energy that they can use. They do not create new energy. When energy is transformed, some nearly always becomes heat. Heat transfers between materials easily, from warmer objects to cooler ones. If no more heat is added, eventually all of a material will reach the same temperature.

Temperature

Temperature is a measure of how fast the atoms in a material are vibrating. High temperature particles vibrate faster than low temperature particles. Rapidly vibrating atoms smash together, which generates heat. As a material cools down, the atoms vibrate more slowly and collide less frequently. As a result, they emit less heat.

- Temperature measures how fast a material's atoms are vibrating.
- Heat measures the material's total energy.

Which has higher heat and which has higher temperature: a candle flame or a bathtub full of hot water?

- The flame has higher temperature, but less heat, because the hot region is very small.
- The bathtub has lower temperature but contains much more heat because it has many more vibrating atoms. The bathtub has greater total energy.

Heat

Heat is taken in or released when an object changes state, or changes from a gas to a liquid, or a liquid to a solid. This heat is called latent heat. When a substance changes state, latent heat is released or absorbed. A substance that is changing its state of matter does not change temperature. All of the energy that is released or absorbed goes toward changing the material's state.

For example, imagine a pot of boiling water on a stove burner: that water is at 100 degrees C (212 degrees F). If you increase the temperature of the burner, more heat enters the water. The water remains at its boiling temperature, but the additional energy goes into changing the water from liquid to gas. With more heat the water evaporates more rapidly. When water changes from a liquid to a gas it takes in heat. Since evaporation takes in heat, this is called evaporative cooling. Evaporative cooling is an inexpensive way to cool homes in hot, dry areas.

Substances also differ in their specific heat, the amount of energy needed to raise the temperature of one gram of the material by 1.0 degrees C (1.8 degrees F). Water has a very high specific heat, which means it takes a lot of energy to change the temperature of water. Let's compare a puddle and asphalt, for example. If you are walking barefoot on a sunny day, which would you rather walk across, the shallow puddle or an asphalt parking lot? Because of its high specific heat, the water stays cooler than the asphalt, even though it receives the same amount of solar radiation.

Energy from the Sun

The earth constantly tries to maintain an energy balance with the atmosphere. Most of the energy that reaches the Earth's surface comes from the Sun. About 44% of solar radiation is in the visible light wavelengths, but the Sun also emits infrared, ultraviolet, and other wavelengths. When viewed together, all of the wavelengths of visible light appear white. But a prism or water droplets can break the white light into different wavelengths so that separate colors appear.

Of the solar energy that reaches the outer atmosphere, UV wavelengths have the greatest energy. Only about 7 percent of solar radiation is in the UV wavelengths. The three types are:

- UVC: The highest energy ultraviolet, does not reach the planet's surface at all.
- UVB: The second highest energy, is also mostly stopped in the atmosphere.
- UVA: The lowest energy, travels through the atmosphere to the ground.

The remaining solar radiation is the longest wavelength, infrared. Most objects radiate infrared energy, which we feel as heat. Some of the wavelengths of solar radiation traveling through the atmosphere may be lost because they are absorbed by various gases. Ozone completely removes UVC, most UVB and some UVA from incoming sunlight. Oxygen, carbon dioxide, and water vapor also filter out some wavelengths.

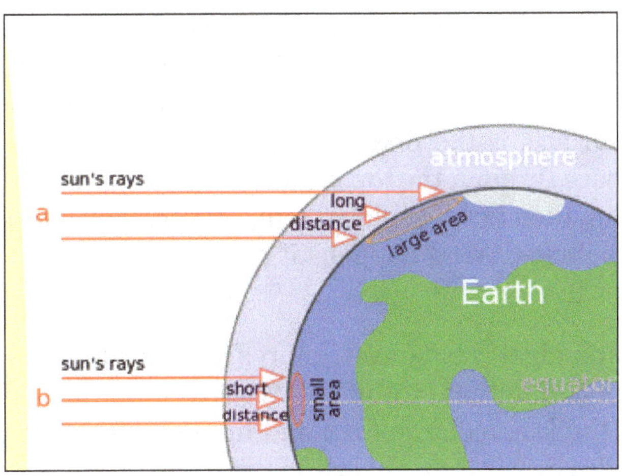

Heat Transfer in the Atmosphere

Heat moves in the atmosphere the same way it moves through the solid Earth or another medium. What follows is a review of the way heat flows and is transferred, but applied to the atmosphere.

Radiation is the transfer of energy between two objects by electromagnetic waves. Heat radiates from the ground into the lower atmosphere.

In conduction, heat moves from areas of more heat to areas of less heat by direct contact. Warmer molecules vibrate rapidly and collide with other nearby molecules, transferring their energy. In the atmosphere, conduction is more effective at lower altitudes where air density is higher; transfers heat upward to where the molecules are spread further apart or transfers heat laterally from a warmer to a cooler spot, where the molecules are moving less vigorously.

Heat transfer by movement of heated materials is called convection. Heat that radiates from the ground initiates convection cells in the atmosphere.

Heat at Earth's Surface

About half of the solar radiation that strikes the top of the atmosphere is filtered out before it reaches the ground. This energy can be absorbed by atmospheric gases, reflected by clouds, or scattered. Scattering occurs when a light wave strikes a particle and bounces off in some other direction.

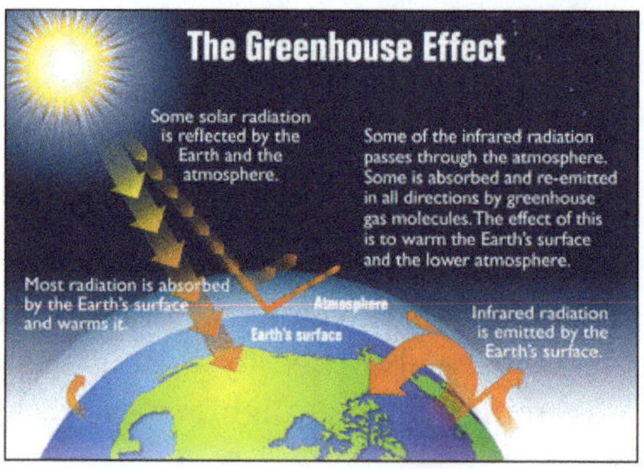

About 3% of the energy that strikes the ground is reflected back into the atmosphere. The rest is absorbed by rocks, soil, and water and then radiated back into the air as heat. These infrared wavelengths can only be seen by infrared sensors. Because solar energy continually enters Earth's atmosphere and ground surface, is the planet getting hotter? The answer is no because energy from Earth escapes into space through the top of the atmosphere. If the amount that exits is equal to the amount that comes in, then average global temperature stays the same. This means that the planet's heat budget is in balance. What happens if more energy comes in than goes out? If more energy goes out than comes in?

To say that the Earth's heat budget is balanced ignores an important point. The amount of incoming solar energy is different at different latitudes). Where do you think the most solar energy ends up and why? Where does the least solar energy end up and why? The difference in solar energy received at different latitudes drives atmospheric circulation.

	Day Length	Sun Angle	Solar Radiation	Albedo
Equatorial Regions	Nearly same all year	High	High	Low
Polar Regions	Night 6 months	Low	Low	High

The Greenhouse Effect

The exception to Earth's temperature being in balance is caused by greenhouse gases. But first the role of greenhouse gases in the atmosphere must be explained. Greenhouse gases warm the atmosphere by trapping heat. Some of the heat radiation out from the ground is trapped by greenhouse gases in the troposphere. Like a blanket on a sleeping person, greenhouse gases act as insulation for the planet. The warming of the atmosphere because of insulation by greenhouse gases is called the greenhouse effect. Greenhouse gases are the component of the atmosphere that moderate Earth's temperatures.Greenhouse gases include CO_2, H_2O, methane, O_3, nitrous oxides (NO and NO_2), and chlorofluorocarbons (CFCs). All are a normal part of the atmosphere except CFCs. The table below shows how each greenhouse gas naturally enters the atmosphere.

Greenhouse Gas	Where It Comes From
Carbon dioxide (CO_2)	Respiration, volcanic eruptions, decomposition of plant material; burning of fossil fuels
Methane	Decomposition of plant material under some conditions, biochemical reactions in stomachs
Nitrous oxide	Produced by bacteria; burning fossil fuels
Ozone	Atmospheric processes, chemical reactions resulting from burning fossil fuels
Chlorofluorocarbons (CFC)	Not naturally occurring; made by humans

Different greenhouse gases have different abilities to trap heat. For example, one methane molecule traps 30 times as much heat as one CO_2 molecule. One CFC-12 molecule (a type of CFC) traps 10,600 times as much heat as one CO_2. Still, CO_2 is a very important greenhouse gas because it is much more abundant in the atmosphere.

Human activity has significantly raised the levels of many of greenhouse gases in the atmosphere. Methane levels are about 2 1/2 times higher as a result of human activity. Carbon dioxide has increased more than 35%. CFCs have only recently existed.

More greenhouse gases trap more heat and warm the atmosphere. The increase or decrease of greenhouse gases in the atmosphere affect climate and weather the world over.

ATMOSPHERIC SCIENCE

Atmospheric science is the interdisciplinary field of study that combines the components of physics and chemistry that focus on the structure and dynamics of Earth's atmosphere. Mathematical tools, such as differential equations and vector analysis, and computer systems are used to evaluate the physical and chemical relations that describe the workings of the atmosphere.

The atmospheric sciences are traditionally divided into three topical areas—meteorology (the study and forecasting of weather), climatology (the study of long-term atmospheric patterns and their influences), and aeronomy (the study of the physics and chemistry of the upper atmosphere). In meteorology, the focus of study concerns day-to-day and hour-to-hour changes in weather within the lower stratosphere and troposphere. Climatology, on the other hand, concentrates more on longer time periods ranging from a single month to millions of years and attempts to describe the interaction of the atmosphere with the oceans, lakes, land, and glaciers. For example, of the three topical areas, climatology would be the best equipped to provide a farmer with the most likely date of the first frost in the autumn. The focus of aeronomy is on the atmosphere from the stratosphere outward. This field also considers the role the atmosphere plays in the propagation of electromagnetic communications, such as shortwave radio transmissions.

Within these three major topical areas, the broad nature of the atmospheric sciences has spawned practitioners who specialize in several distinct subfields. Scientists who investigate the physics associated with atmospheric flow are called dynamic meteorologists or simply dynamicists. When the investigation procedure involves the application of large computer models of atmospheric structure and dynamics, the scientists are called numerical modelers. Scientists and technicians who specifically investigate procedures of weather forecasting are called synoptic meteorologists, while those who investigate the physical mechanisms associated with the growth of cloud droplets and ice crystals and related precipitation processes are called cloud physicists. Researchers who study atmospheric optical effects are referred to as physical meteorologists, while individuals who investigate the dynamics and observations of climate are called climatologists or climate scientists. Paleoclimatologists are researchers who concentrate on ancient climate patterns. Scientists who investigate atmospheric structure and dynamics within the boundary layer (the layer of the atmosphere closest to Earth's surface) are referred to as boundary layer meteorologists or micrometeorologists.

2

Branches of Atmospheric Science

Atmospheric science is a vast discipline that branches into several sub-disciplines. Some of these are climatology, meteorology, atmospheric chemistry, atmospheric dynamics, coniology and atmospheric physics. This chapter has been carefully written to provide an easy understanding of these branches of atmospheric science.

METEOROLOGY

Meteorology is the study of the atmosphere, atmospheric phenomena, and atmospheric effects on our weather. The atmosphere is the gaseous layer of the physical environment that surrounds a planet. Earth's atmosphere is roughly 100 to 125 kilometers (65-75 miles) thick. Gravity keeps the atmosphere from expanding much farther.

Meteorology is a subdiscipline of the atmospheric sciences, a term that covers all studies of the atmosphere. A subdiscipline is a specialized field of study within a broader subject or discipline. Climatology and aeronomy are also subdisciplines of the atmospheric sciences. Climatology focuses on how atmospheric changes define and alter the world's climates. Aeronomy is the study of the upper parts of the atmosphere, where unique chemical and physical processes occur. Meteorology focuses on the lower parts of the atmosphere, primarily the troposphere, where most weather takes place.

Meteorologists use scientific principles to observe, explain, and forecast our weather. They often focus on atmospheric research or operational weather forecasting. Research meteorologists cover several subdisciplines of meteorology to include: climate modeling, remote sensing, air quality, atmospheric physics, and climate change. They also research the relationship between the atmosphere and Earth's climates, oceans, and biological life.

Forecasters use that research, along with atmospheric data, to scientifically assess the current state of the atmosphere and make predictions of its future state. Atmospheric conditions both at the Earth's surface and above are measured from a variety of sources: weather stations, ships, buoys, aircraft, radar, weather balloons, and satellites. This data is transmitted to centers throughout the world that produce computer analyses of global weather. The analyses are passed on to national and regional weather centers, which feed this data into computers that model the future state of the atmosphere. This transfer of information demonstrates how weather and the study of it take place in multiple, interconnected ways.

Scales of Meteorology

Weather occurs at different scales of space and time. The four meteorological scales are: microscale, mesoscale, synoptic scale, and global scale. Meteorologists often focus on a specific scale in their work.

Microscale Meteorology

Microscale meteorology focuses on phenomena that range in size from a few centimeters to a few kilometers, and that have short life spans (less than a day). These phenomena affect very small geographic areas, and the temperatures and terrains of those areas.

Microscale meteorologists often study the processes that occur between soil, vegetation, and surface water near ground level. They measure the transfer of heat, gas, and liquid between these surfaces. Microscale meteorology often involves the study of chemistry.

Tracking air pollutants is an example of microscale meteorology. MIRAGE-Mexico is a collaboration between meteorologists in the United States and Mexico. The program studies the chemical and physical transformations of gases and aerosols in the pollution surrounding Mexico City. MIRAGE-Mexico uses observations from ground stations, aircraft, and satellites to track pollutants.

Mesoscale Meteorology

Mesoscale phenomena range in size from a few kilometers to roughly 1,000 kilometers (620 miles). Two important phenomena are mesoscale convective complexes (MCC) and mesoscale convective systems (MCS). Both are caused by convection, an important meteorological principle.

Convection is a process of circulation. Warmer, less-dense fluid rises, and colder, denser fluid sinks. The fluid that most meteorologists study is air. (Any substance that flows is considered a fluid.) Convection results in a transfer of energy, heat, and moisture—the basic building blocks of weather.

In both an MCC and MCS, a large area of air and moisture is warmed during the middle of the day—when the sun angle is at its highest. As this warm air mass rises into the colder atmosphere, it condenses into clouds, turning water vapor into precipitation.

An MCC is a single system of clouds that can reach the size of the state of Ohio and produce heavy rainfall and flooding. An MCS is a smaller cluster of thunderstorms that lasts for several hours. Both react to unique transfers of energy, heat, and moisture caused by convection.

The Deep Convective Clouds and Chemistry (DC3) field campaign is a program that will study storms and thunderclouds in Colorado, Alabama, and Oklahoma. This project will consider how convection influences the formation and movement of storms, including the development of lightning. It will also study their impact on aircraft and flight patterns. The DC3 program will use data gathered from research aircraft able to fly over the tops of storms.

Synoptic Scale Meteorology

Synoptic-scale phenomena cover an area of several hundred or even thousands of kilometers. High- and low-pressure systems seen on local weather forecasts, are synoptic in scale. Pressure, much like convection, is an important meteorological principle that is at the root of large-scale weather systems as diverse as hurricanes and bitter cold outbreaks.

Low-pressure systems occur where the atmospheric pressure at the surface of the Earth is less than its surrounding environment. Wind and moisture from areas with higher pressure seek low-pressure systems. This movement, in conjunction with the Coriolis force and friction, causes the system to rotate counter-clockwise in the Northern Hemisphere and clockwise in the Southern Hemisphere, creating a cyclone. Cyclones have a tendency for upward vertical motion. This allows moist air from the surrounding area to rise, expand and condense into water vapor, forming clouds. This movement of moisture and air causes the majority of our weather events.

Hurricanes are a result of low-pressure systems (cyclones) developing over tropical waters in the Western Hemisphere. The system sucks up massive amounts of warm moisture from the sea, causing convection to take place, which in turn causes wind speeds to increase and pressure to fall. When these winds reach speeds over 119 kilometers per hour (74 miles per hour), the cyclone is classified as a hurricane.

Hurricanes can be one of the most devastating natural disasters in the Western Hemisphere. The National Hurricane Center, in Miami, Florida, regularly issues forecasts and reports on all tropical weather systems. During hurricane season, hurricane specialists issue forecasts and warnings for every tropical storm in the western tropical Atlantic and eastern tropical Pacific. Businesses and government officials from the United States, the Caribbean, Central America, and South America rely on forecasts from the National Hurricane Center.

High-pressure systems occur where the atmospheric pressure at the surface of the Earth is greater than its surrounding environment. This pressure has a tendency for downward vertical motion, allowing for dry air and clear skies.

Extremely cold temperatures are a result of high-pressure systems that develop over the Arctic and move over the Northern Hemisphere. Arctic air is very cold because it develops over ice and snow-covered ground. This cold air is so dense that it pushes against Earth's surface with extreme pressure, preventing any moisture or heat from staying within the system.

Meteorologists have identified many semi-permanent areas of high-pressure. The Azores high, for instance, is a relatively stable region of high pressure around the Azores, an archipelago in the mid-Atlantic Ocean. The Azores high is responsible for arid temperatures of the Mediterranean basin, as well as summer heat waves in Western Europe.

Global Scale Meteorology

Global scale phenomena are weather patterns related to the transport of heat, wind, and moisture from the tropics to the poles. An important pattern is global atmospheric circulation, the large-scale movement of air that helps distribute thermal energy (heat) across the surface of the Earth.

Global atmospheric circulation is the fairly constant movement of winds across the globe. Winds develop as air masses move from areas of high pressure to areas of low pressure. Global atmospheric circulation is largely driven by Hadley cells. Hadley cells are tropical and equatorial convection patterns. Convection drives warm air high in the atmosphere, while cool, dense air pushes lower in a constant loop. Each loop is a Hadley cell.

Hadley cells determine the flow of trade winds, which meteorologists forecast. Businesses, especially those exporting products across oceans, pay close attention to the strength of trade winds because they help ships travel faster. Westerlies are winds that blow from the west in the midlatitudes. Closer to the Equator, trade winds blow from the northeast (north of the Equator) and the southeast (south of the Equator).

Meteorologists study long-term climate patterns that disrupt global atmospheric circulation. Meteorologists discovered the pattern of El Nino, for instance. El Niño involves ocean currents and trade winds across the Pacific Ocean. El Niño occurs roughly every five years, disrupting global atmospheric circulation and affecting local weather and economies from Australia to Peru.

El Niño is linked with changes in air pressure in the Pacific Ocean known as the Southern Oscillation. Air pressure drops over the eastern Pacific, near the coast of the Americas, while air pressure rises over the western Pacific, near the coasts of Australia and Indonesia. Trade winds weaken. Eastern Pacific nations experience extreme rainfall. Warm ocean currents reduce fish stocks, which depend on nutrient-rich upwelling of cold water to thrive. Western Pacific nations experience drought, devastating agricultural production.

Understanding the meteorological processes of El Niño helps farmers, fishers, and coastal residents prepare for the climate pattern.

Meteorology Today

Today's meteorologists have a variety of tools that help them examine, describe, model, and predict weather systems. These technologies are being applied at different meteorological scales, improving forecast accuracy and efficiency.

Radar is an important remote sensing technology used in forecasting. A radar dish is an active sensor in that it sends out radio waves that bounce off particles in the atmosphere and return to the dish. A computer processes these pulses and determines the horizontal dimension of clouds and precipitation, and the speed and direction in which these clouds are moving.

A new technology, known as dual-polarization radar, transmits both horizontal and vertical radio wave pulses. With this additional pulse, dual-polarization radar is better able to estimate precipitation. It is also better able to differentiate types of precipitation—rain, snow, sleet, or hail. Dual-polarization radar will greatly improve flash-flood and winter-weather forecasts.

Tornado research is another important component of meteorology. Starting in 2009, the National Oceanic and Atmospheric Administration (NOAA) and the National Science Foundation conducted the largest tornado research project in history, known as VORTEX2. The VORTEX2 team, consisting of about 200 people and more than 80 weather instruments, traveled more than 16,000 kilometers (10,000 miles) across the Great Plains of the United States to collect data on how, when, and why tornadoes form. The team made history by collecting extremely detailed data before, during, and after a specific tornado. This tornado is the most intensely examined in history and will provide key insights into tornado dynamics.

Satellites are extremely important to our understanding of global scale weather phenomena. The National Aeronautics and Space Administration (NASA) and NOAA operate three Geostationary Operational Environmental Satellites (GOES) that provide weather observations for more than 50 percent of the Earth's surface.

GOES-15, launched in 2010, includes a solar X-ray imager that monitors the sun's X-rays for the early detection of solar phenomena, such as solar flares. Solar flares can affect military and commercial satellite communications around the globe. A highly accurate imager produces visible and infrared images of Earth's surface, oceans, cloud cover, and severe storm developments. Infrared imagery detects the movement and transfer of heat, improving our understanding of the global energy balance and processes such as global warming, convection, and severe weather.

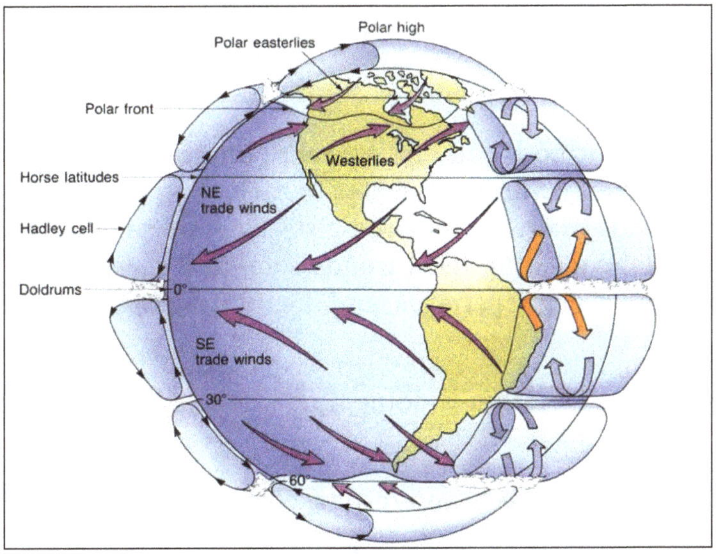

Hadley cells are the basis for our understanding of global-scale meteorology.

CLIMATOLOGY

Climatology, or sometimes known as climate science, is the study of the Earth's weather patterns and the systems that cause them. From the ocean oscillations to trade winds, pressure systems that drives temperature, airborne particles that influence local conditions and even the phases of the moon and Earth's wobble all affect the climate. This means that climatology is the "study of zones" although in reality it is much more complicated than that.

Climatologists today are almost universally directing their efforts towards understanding, explaining and attempting to do something about global warming, but as you will see in the list below, that is not the science's only puzzle to solve nor the limits of its interest. Until relatively recently, it was considered a dry and uninteresting, yet necessary area of science. But since it became clear that human actions are damaging the environment and changing the climate, it has become much more prominent nationally and internationally with most government departments in most countries having responsibilities to mitigate or prepare for climate change scenario.

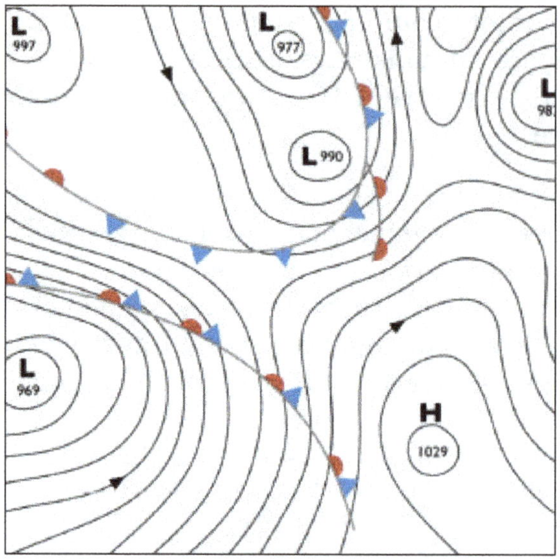

Climatology is not meteorology; both concern weather patterns and their causes and effects, but differ in many ways although they will be interested in each other's data. Climate science examines long-term patterns and trends whereas meteorologists examine short-term weather patterns and their impact, climatologists study long-term trends such as temperature change, water and ice levels, cloud cover, flood and drought patterns, and their long-term, long-range impact on various topographies and globally.

- Meteorology is what the weather is doing now whereas climatology is what you expect to see.
- Meteorology is short-term effects and results, climatology is long-term consequences.
- Meteorology concerns small areas; climatology concerns much larger areas, or global results.

Even how they use equipment differs. Satellites, for example, are used in meteorology to track weather systems and to monitor atmospheric fronts to predict what the weather will do next. Climatologists use different data sets from satellites - temperatures at various levels through the atmosphere and at ground level, for example, or mapping energy flows and noticing changes that could affect local climate.

Natural Variables that Influence our Climate

Oscillations

Much of the discussion of climatology in public discourse concerns anthropogenic climate change - the contribution of human activity to such events as carbon particles, greenhouse gases, and their effects such as the Greenhouse Effect and coral bleaching. But climatology is not just about the human impact, it is about looking at and predicting the effects of the natural processes on our planet and the Solar System. It is to long-term trends, long-term and large-scale processes to which we direct our efforts in understanding why the climate looks or acts the way it does, and what variations and anomalies can potentially change it. Here are some of the natural trends.

El Niño and La Niña Oscillation

Perhaps the two best-known oscillations are El Niño and La Niña. They occur roughly once every seven years, and the latter always follows the former after a few months. They are usually predictable, but the increasing effects of climate change can sometimes push them back to a little later. Known as ENSO, they are opposite effects of the same process and are defined as an oscillation (a variation in magnitude) between the temperature of the atmosphere and the ocean. In this case, ENSO occurs in the eastern equatorial Pacific. Geographically speaking, it approximates to the area between the International Dateline and 120 degrees west of there. El Niño typically arrives between June and December in a given year and takes place with the depletion and failure of replacement of the Pacific Trade Winds following the Pacific monsoon season. The warm air then creates this oscillation and the waters become warmer.

La Niña occurs when the situation is reversed; it usually follows El Niño within a year. Here, trade winds are stronger than normal, moving warm water westward across the Pacific. The east of the ocean is colder than normal while the west is warmer than average. But the effects don't just affect the Pacific, they are global.

The Effects of ENSO

The biggest impacts are felt in the Pacific area in the immediate and short term, but none of these climate systems exists in isolation. They create knock-on effects elsewhere.

With El Niño, Peruvian fishermen find that anchovy stocks are low. Precipitation shifts globally. The Iberian Peninsula in Europe (Spain and Portugal) tends to be wetter from August to November, the tip of South America tends to have higher rainfall too while tropical Africa is drier than normal. But precipitation is not the only impact. Northern Europe tends to be much colder in the winter months while central and western Europe will experience warmer than normal temperatures in the autumn. Here in the US, the Eastern Seaboard will experience lower precipitation between December and May while the same area will experience colder temperatures between June and September - the height of the summer months.

With La Niña, these effects will tend to be flipped. Central Mexico and the southern US can expect lower precipitation between the following December and March while the opposite is true just a few thousand miles to the south in northern areas of South America. In fact, most of the central tropics from Africa through the Indian Ocean to the western Pacific will experience higher than average rainfall.

Madden-Julian Oscillation

Not as well known as ENSO, the Madden-Julian Oscillation (MJO), it functions differently. While ENSO is static, occurring in one place due to the causes that create those conditions, MJO moves across a geographical area - eastwards along the tropics, bringing higher levels of cloud, increased precipitation and therefore the risk of rainfall. It returns to its starting position anything between one and two months after it began. It can be tracked and was only identified in the 1970s. There may be multiple events in a single place since it moves. It also experiences variable speed, creating conditions ahead, during, and behind it. With each event, there is an increased rainfall event and a suppressed rainfall event.

Tropical weather events are predictable once MJO is on the move, this means preparing for extreme weather (potential drought and flooding) in the tropical areas. But it can have much longer range effects, creating knock-on effects as far away as the northern US, Canada, and northern Europe. This is why weather services in the northern hemisphere are increasingly interested in incorporating its effects into their monitoring and prediction services.

North Atlantic Oscillation

The North Atlantic Oscillation (or the NAO) occurs in the northern area of the Atlantic Ocean, the second largest ocean on our planet. It describes fluctuations between the sea level atmospheric pressure from the areas known as the Azores High and the Icelandic Low although these are not set in stone. When the waters fluctuate between these two points, the westerly winds direction and strength change. In turn, this also affects the strength, frequency and direction of Atlantic storms. It has been known since the 19th century, and unlike some of the others discussed in this part, it is mostly atmospheric, affected and part of the Arctic Oscillations.

It is a natural oscillation and can massively impact the continents on either side of it - North America and Europe. When one side has certain conditions, the other side has opposite conditions. It will also affect the North Atlantic Jet Stream which keeps Western Europe and North America temperate compared to adjacent areas.

Positive NAO is when warm westerly winds dominate the North Atlantic, bringing warm air from the southern US coastal states northeast towards northwestern Europe. The British Isles, for example, tend to experience warm but wet weather. However, this direction means colder weather in the northern coastal states and Canada on the Atlantic side and southwestern Europe.

Negative NAO also carries weather fronts from North America to Western Europe but in this case, the southern US states experience cold and snowy weather conditions while their northern counterparts experience warmer weather. In Europe, Scandinavia tends to be cold but dry while the Mediterranean Basin is warmer and wetter.

North Pacific Oscillation

The North Pacific Oscillation works similarly to the North Atlantic Oscillation "seesaw" system divided by northern and southern regions. Also, like the NAO, there are two phases. The first is the Aleutian Below (south of the Aleutian Islands), the other the Aleutian Above (north of the islands). They create variable weather patterns depending on which is currently in progress. It's not a well-known oscillation as far as the public are concerned, but its impacts are far greater than the NAO and ENSO. It runs from roughly Alaska to Hawaii.

During the "above" phase (known as the deeper Aleutian low), the North American continent is warmer except for certain areas of the western US with higher levels of precipitation over the Pacific Northwest. There is also impact on sea ice during the positive NPO with coverage being more extensive in Bering and Okhotsk Seas.

Pacific Decadal Oscillation

Similar to the ENSO, the PDO occurs over much longer timeframes - measured in decades, hence the name. Typically, the phase lasts for around 20 or 30 years compared to ENSO which lasts no more than six years (incorporating both El Niño and La Niña). It also incorporates both a warm and a cool period during the oscillation - both of which impact upper atmospheric winds. Despite its geographic limitations, it can impact global weather and climate, including frequency and intensity of hurricanes across the Pacific and even in the Atlantic. Sometimes, hurricanes in the western US and into Europe are a result of what happened in the Pacific, impacting both flooding and drought effects. Land temperatures and marine ecosystems are affected - the former being increased while the latter's production of plankton reduces, impacting the wider ecosystem. It also has an interesting relationship with ENSO. When they occur together, impacts are magnified but when they are out of phase, it appears effects are effectively mitigated.

Interdecadal Pacific Oscillation

The Interdecadal Pacific Oscillation (IPO) is similar to the PDO but it covers the area of the Pacific area south of the PDO (south of the 20 degrees N latitude), and lasts anything from 15 to 30 years. When the IPO is positive, temperatures are lower in the Pacific tropical and subtropical areas but warmer farther to the north. Also, like the PDO, it can have amplification or mitigation effects in conjunction with ENSO with particular and noteworthy differences to the climate of Australia, New Zealand and the seas around those countries.

Other Natural Climate Drivers

The oscillations are just some of the natural forcings on the climate. There are anthropogenic forcings, but the wind and air interactions are not the only naturally-occurring events.

Sun Activity

At the center of our solar system is a star that we call "The Sun". The sun receives a constant stream of energy produced by nuclear processes that generate an immense amount of heat, light, and radiation. This radiation hits the atmosphere and is absorbed by the ozone layer which reduces the most harmful. The resulting heat and ultraviolet rays that do make it to the surface enable all life on the planet from the plants that use the chemical processes to reproduce through photosynthesis which, in turn, feeds animals. But the sun's energy is not constant. The temperature and therefore the heat levels fluctuate and that can have a knock-on effect for the Earth. It seems that the Sun was responsible for two climate events that took place less than 1,000 years ago: the Little Ice Age and the Medieval Warm Period. The decrease in solar temperature certainly contributed to the LIA, but it was by no means the only factor - volcanoes also impacted global temperatures - there was a lot of it in the Little Ice Age. The Medieval Warm Period (MWP) certainly correlates with an increase in solar activity and a decrease in volcanic activity during the period.

As atmospheric particles increase, so does air density. This leads to the greenhouse effect. Less heat is reflected back into space and more is absorbed in the upper atmosphere. This is what happened on the planet Venus which has dense cloud cover and a runaway Greenhouse Effect. Mars atmosphere is too thin; when this happens, too much heat escape, leading to the opposite effect.

Volcanoes

These are immense seismological events that can cause short-term weather effects and, when enough events have accumulated or are particularly large, long-term climate effects. It is largely believed that the explosion of the Thera (Santorini) volcano in antiquity led to the collapse of several complex societies, not just the Minoan civilization that inhabited the island at the time, but also deeply affected other powers too and led to the downfall of the Minoan civilization which, just a few years later, was conquered by the Mycenaeans. Writings about the event are found all over the world with some writings from China and physical effects on the climate discovered in North America. It was five times larger than Krakatoa and no other volcano has come close to matching its power. The turmoil that followed was political but also showed the lack of resources in societies led to their collapse.

The Little Ice Age created a brief yet intense change to the northern hemisphere's climate. This was a period of unusually high volcanic activity but evidence on whether this was a cause is still limited and subject to much speculation. But we do know that volcanic activity has an impact - mostly small, but sometimes when there is a lot together or there is a large enough single event, change occurs. In more recent times, Krakatoa in the East Indies cooled the planet's temperature rise and halted sea level rise for at least two decades, and that was one volcano - albeit an intense one.

Sub Divisions of Climatology

Climatology in its modern form is less than a century old, but already a number of subdivisions have grown up within the discipline to cope with the data and to create niches for experts to specialize.

Applied Climatology

As with applied chemistry, applied physics and so on, applied climatology is about studying what is actually happening now rather than climate theory of what will happen, or the use of theoretical models to predict events in the short or long-term future. This means that applied climatology has much in common with some of the other atmospheric sciences such as meteorology. Climate events that are happening now (for example, any of the oscillations mentioned above) have immediate and measurable impacts on weather systems - locally and far away. It is arguably the oldest form of climatology, originating long before the birth of the modern science. Military decisions were made based on climate events and weather patterns, and Ancient Egypt's system of agriculture was based on the observation of the fluctuations in the Nile's "inundation" (flood waters bringing nutrient-rich silt that was the backbone of their agricultural surplus) allowed for long-term planning of social, agricultural and civic project planning.

Bioclimatology

Overlapping with ecology in many ways, bioclimatology is the study of climate change (both natural and anthropogenic) on various life forms and ecological systems. It provided one of the earliest conundrums for naturalists, especially with the discovery of ancient aquatic fossils in desert areas, and fossilized tropical botanical remains found in areas such as tundra. As a climate changes - getting wetter or drier, warmer or cooler, that will have a measurable and noticeable impact on the biodiversity as well as the species profiles within these areas. Mostly, bioclimatology researchers are concerned with the human impact on ecology and biodiversity, but they study any landscape change resulting from climate change. Some researchers will also overlap with medical research. As conditions north of the equator in Eurasia and the Americas become warmer and humidity increases, we may observe increasing instances of malarial mosquitos into northern Mexico, the southern US and southern Europe, for example.

Boundary-layer Climatology

The Boundary Layer is the lowest level of the atmosphere, the area most affected by local and planetary climate change. This is the atmospheric layer that experiences the most turbulence and holds the important weather systems, interacting with the "thermals", distributing air and moisture across the planet. Therefore, boundary layer climatologists study the networks and interactions of the lower atmosphere, and their impact on weather and climate systems, but also how the network of winds are affected by such climate phenomena as rising air and sea temperatures, urban heat islands, and natural events such as volcanic activity. It also has critical uses for immediate weather patterns in meteorology. Neither boundary-layer climatology nor meteorology exist in isolation and are intrinsically linked.

Dynamic Climatology

No science exists in a bubble and no subdivision of a broad science does so either. Dynamic climatologists are concerned with examining the accumulated sum total of information acquired from all related sciences, typically quantitative in nature, based on observed phenomena. Dynamic climatology examines and handles everything from paleodata to volcanic eruption to looking at short range or short-term weather patterns to long-term climate effects from natural or anthropogenic causes. This area of climatology uses a holistic approach.

Historical Climatology

While paleoclimatology is concerned with the climate of the ancient past, historical climatology is concerned with climate change, alteration and patterns that have existed on the human measurement scale. It blends climatology with environmental history and sometimes, can complement research in the human Earth sciences such as anthropology, archaeology and human geography. It seeks to recreate or profile past environments, examining past natural disasters and their effects. This would include the study of the Little Ice Age and the Medieval Warm Period, the environmental changes that resulted from volcanic explosions at Krakatoa and Thira. However, it can also examine the human impact on the environment and the after-effects of such human activities as land clearance for agriculture (for example, the Neolithic Agricultural Revolution shows clear and distinct topographical change and is traceable in the ecological and archaeological record).

Hydroclimatology

How do the natural and anthropogenic changes to the climate affect our waterways? Increased drought and flooding, ocean acidification, coral bleaching, the effects of ocean temperature and pH of our oceans affect how much plankton is produced in any cycle. In turn, this affects the ocean lifecycle. Hydroclimatology is a vitally important area of climatology because around 2/3 of our planet is covered in ocean. It's a vast, complex and vital ecosystem. Sea level rises, ice cap melt, ocean particles, the oceans as a carbon sink, precipitation and drought all contribute to something called the "Water Budget" and it is this that hydroclimatologists spend most of their time analyzing.

Paleoclimatology

Historical climatology covers the historical record - starting with the invention of instruments able to take measurements while paleoclimatology is concerned with the entire history of the planet's climate record Typical data sets include taking radiocarbon dates and chemical signatures as ecological indicators from such areas as tree rings - also known as dendrochronology - Antarctic and Arctic ice cores, information from fossilized coral and vegetation, and lake and river sediments. It is through this area that we know the Earth has always changed; it is how we know that Earth has undergone at least five ice ages, the Medieval Warm Period and Little Ice Age, and, when comparing ecological data and fossils, how we know the effects that high or low atmospheric carbon will have on the climate. It has also been integral to demonstrating how quickly a climate can change - with a catatrophe or massive climatic event, sometimes within a matter of decades.

Physical Climatology

Most of climatology is concerned with looking at data and making projections or presenting the data as facts, statistics, graphs and hard figures - it is quantitative in nature. Physical climatology is more qualitative. It examines and explains how climate can shape topography and geographical systems. For example, it seeks to explain how glaciation is one factor capable of forming valleys and mountains, how extreme flooding events will change a landscape. For example, we know that the Vlad - of North America's largest glaciers created the enormous basin around the Great Lakes and then filled it with meltwater to create the lakes we have today.

Synoptic Climatology

This branch of climatology is concerned with circulation patterns within the atmosphere, paying particular attention to how these circulations create differences in climate between either topographically comparable or geographically close locations. It creates categories of synoptic climate patterns and then attempts to discern what a climate is going to look like in the immediate future based on season weather patterns and anomalous phenomena. Typically, they may be concerned with the weather patterns that create hurricanes and tornadoes.

The Human Impacts on Climate

Of course, climatology is not just about the natural variations in the short, medium or long term, it's primarily concerned today with the human impact on climate change and the various "forcings" that are already causing problems. Most of the following issues have impacted the climate since the Industrial Revolution in the 19th century and arguably started before that.

Greenhouse Gases

Greenhouse gases are so-called because their abundance leads to a "greenhouse effect". Greenhouses are glass buildings used to houseplants that require humidity and shelter from the elements. By nature, they are warmer and more humid than a garden. The effect on the global environment is similar. The greenhouse gases are methane, water vapor, nitrous oxide and carbon dioxide.

- Water Vapor creates a feedback - precipitation and cloud cover increases, leading to higher temperatures, but also creates more rainfall which can cool surfaces but can lead to flooding in some areas.
- Nitrous Oxide is a byproduct of agricultural processes. It received much attention in the 1980s with so-called acid rain, but the reduction in this does not mean N_2O is no longer a threat.
- Methane is a natural gas and one burnt as fuel in many parts of the world, but it too increases air density and the greenhouse effect. The impact is proportionally high considering the low levels of the gas in the atmosphere.
- Carbon dioxide is the best-known and most important of all greenhouse gases, it is released through several natural processes but also industrial actions - one of the biggest impacting factors since the Industrial Revolution.
- Chlorofluorocarbons (CFCs) are now heavily regulated since it was discovered how much their release was damaging the ozone layer - the gas level protecting the planet from the sun's most harmful rays.

Deforestation

Cutting down tree canopies without replacing it passively increases climate change by the simple fact that trees and other vegetation are carbon sinks. The fewer sinks we have, the faster carbon emissions will accumulate in the atmosphere and it seems deforestation is increasing despite international efforts to slow it down and replace more trees than we are cutting down. Some of this carbon is heading for the oceans which is now absorbing much more carbon than it has done for a very long time. This is leading to ocean acidification and coral bleaching which is upsetting the delicate balance of marine ecosystems, reducing ocean life that has come to rely on coral reefs to survive.

The Future Challenges for Climatology

Many challenges associated with climate change are challenges for ecologists, conservationists, and for politicians and other decision-makers to solve, not necessarily for climate researchers. But climatology does have an exciting future of new discoveries and new technologies; it also has some new challenges to rise to as we head deeper into the 21st century.

Studying the Effects on Landscapes

As climate change brings changes all over the globe, it will be down to climatologists to observe these changes as they happen and continue to publish data for decision-makers to respond. We are already seeing these effects, most notably in the developing world for people who do not have the resources of the first world and live in so-called marginal landscapes. Climatology is at the forefront of looking at these changes and comparing data from different years, decades and even centuries. Some lands are sensitive to changes in the water cycle and impact on aquatic resources most critically - not just for drinking water, but also for crops. Without water, people in the most affected lands cannot feed themselves. Climate change can also mean the difference between whether an area receives rain or snow, and the longer-range and long-term effects including flooding when seasonal temperatures are too high. This has the potential for long-term effects.

Studying the Climate Effects of Urbanism

It can't be halted, either in the developed nor the developing world, but the spread of urbanism in the landscape creates far greater energy consumption than rural land. Homes need heating and lighting, but so do public buildings, street lighting, shopping malls and industry, commerce and residential zones use far greater energy levels per square foot. Even without the resource consumption, the spread of urban development causes intense ecological change. But life adapts, and many lifeforms have thrived in urban centers. Plenty of evidence exists now to suggest that urban centers create microclimates. Urban heat islands create large, warm areas of air that affect regional and even global climate when large enough. This is why urban centers will be a key battleground in mitigating the effects of climate change and why climatologists are so interested in studying their effects over time.

Making Sense of Big Data

While big data has been a welcome addition to the data sets for climatology, the issue remains that quantity does not equal quality. Climatology in the 21st century runs an even greater risk of accumulating useless and irrelevant or even masses of counterproductive anomalous data, and it will be a major challenge for climatologists to further develop their understanding of data quality in this century. New sources include the ever-increasing paleoclimate data, proxy data, and the use of satellite data to take recordings in areas generally inaccessible for manned research - not to mention that using satellite data is far cheaper. Climatology will continue to deliver these data sets and will, in future, be required to provide broader data sets and more accurate analyses for environmental scientists in other areas. More than ever, climatology is beginning to overlap with data science.

Open Access vs. Intellectual Property Rights

Over the last ten years, researchers in many sciences have called for all scientific data to become open access - the idea that anybody, anywhere, with the means to do so, can request the findings of published research. At present, such data is limited to academic journals which are beyond the finances of the average person. Also, governments the fund such data have traditionally been fiercely protective of the data that their public funds have produced. Both paradigms are understandable - governments that invest in public science projects wish for their country to be the sole beneficiary of the results of that data, but that does not help the academic community and it's proving a barrier to both public understanding of climatology and to driving the science forward by arming researchers with as much information as possible.

Paleoclimatology

Paleoclimatology is the study of changes in climate taken on the scale of the entire history of Earth. It uses a variety of proxy methods from the Earth and life sciences to obtain data previously preserved within things such as rocks, sediments, ice sheets, tree rings, corals, shells, and microfossils. It then uses the records to determine the past states of the Earth's various climate regions and its atmospheric system. Studies of past changes in the environment and biodiversity often reflect on the current situation, specifically the impact of climate on mass extinctions and biotic recovery.

Reconstructing Ancient Climates

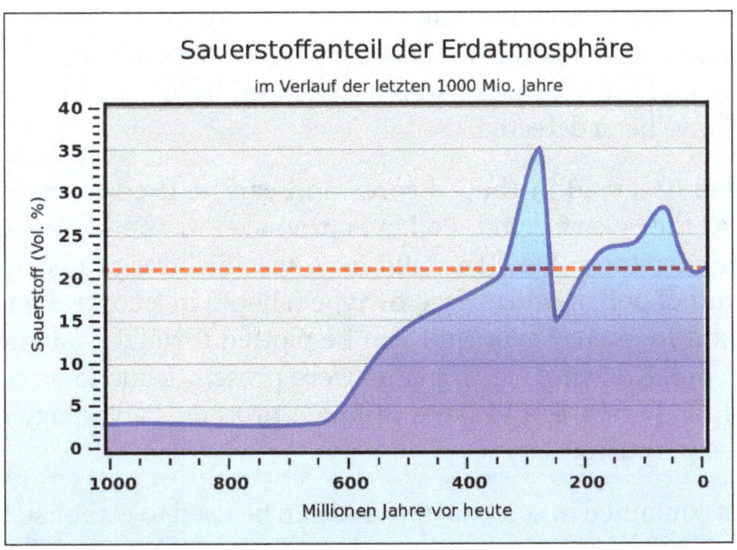

The oxygen content in the atmosphere over the last billion years.

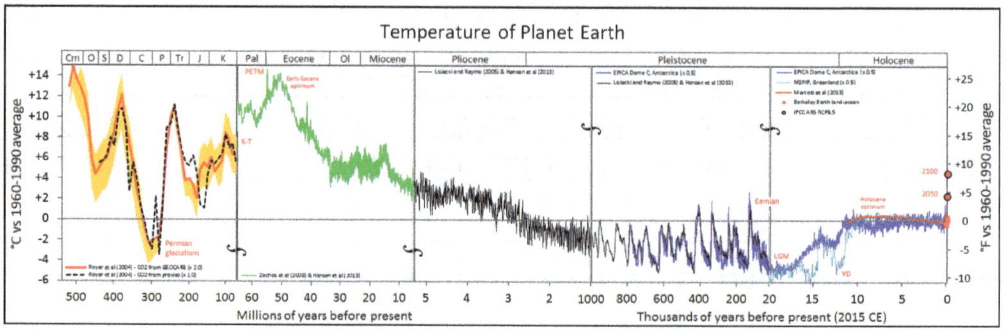

Palaeotemperature graphs compressed together.

Paleoclimatologists employ a wide variety of techniques to deduce ancient climates.

Ice

Mountain glaciers and the polar ice caps/ice sheets provide much data in paleoclimatology. Ice-coring projects in the ice caps of Greenland and Antarctica have yielded data going back several hundred thousand years, over 800,000 years in the case of the EPICA project.

- Air trapped within fallen snow becomes encased in tiny bubbles as the snow is compressed into ice in the glacier under the weight of later years' snow. The trapped air has proven a tremendously valuable source for direct measurement of the composition of air from the time the ice was formed.
- Layering can be observed because of seasonal pauses in ice accumulation and can be used to establish chronology, associating specific depths of the core with ranges of time.
- Changes in the layering thickness can be used to determine changes in precipitation or temperature.
- Oxygen-18 quantity changes ($\delta^{18}O$) in ice layers represent changes in average ocean surface temperature. Water molecules containing the heavier O-18 evaporate at a higher temperature than water molecules containing the normal Oxygen-16 isotope. The ratio of O-18 to O-16 will be higher as temperature increases. It also depends on other factors such as the water's salinity and the volume of water locked up in ice sheets. Various cycles in those isotope ratios have been detected.
- Pollen has been observed in the ice cores and can be used to understand which plants were present as the layer formed. Pollen is produced in abundance and its distribution is typically well understood. A pollen count for a specific layer can be produced by observing the total amount of pollen categorized by type (shape) in a controlled sample of that layer. Changes in plant frequency over time can be plotted through statistical analysis of pollen counts in the core. Knowing which plants were present leads to an understanding of precipitation and temperature, and types of fauna present. Palynology includes the study of pollen for these purposes.
- Volcanic ash is contained in some layers, and can be used to establish the time of the layer's formation. Each volcanic event distributed ash with a unique set of properties (shape and

color of particles, chemical signature). Establishing the ash's source will establish a range of time to associate with layer of ice.

Dendroclimatology

Climatic information can be obtained through an understanding of changes in tree growth. Generally, trees respond to changes in climatic variables by speeding up or slowing down growth, which in turn is generally reflected by a greater or lesser thickness in growth rings. Different species, however, respond to changes in climatic variables in different ways. A tree-ring record is established by compiling information from many living trees in a specific area.

Older intact wood that has escaped decay can extend the time covered by the record by matching the ring depth changes to contemporary specimens. By using that method, some areas have tree-ring records dating back a few thousand years. Older wood not connected to a contemporary record can be dated generally with radiocarbon techniques. A tree-ring record can be used to produce information regarding precipitation, temperature, hydrology, and fire corresponding to a particular area.

Sedimentary Content

On a longer time scale, geologists must refer to the sedimentary record for data.

- Sediments, sometimes lithified to form rock, may contain remnants of preserved vegetation, animals, plankton, or pollen, which may be characteristic of certain climatic zones.
- Biomarker molecules such as the alkenones may yield information about their temperature of formation.
- Chemical signatures, particularly Mg/Ca ratio of calcite in Foraminifera tests, can be used to reconstruct past temperature.
- Isotopic ratios can provide further information. Specifically, the $\delta^{18}O$ record responds to changes in temperature and ice volume, and the $\delta^{13}C$ record reflects a range of factors, which are often difficult to disentangle.

Sea floor core sample labelled to identify the exact spot on the sea floor where the sample was taken. Sediments from nearby locations can show significant differences in chemical and biological composition.

Sedimentary Facies

On a longer time scale, the rock record may show signs of sea level rise and fall, and features such as "fossilised" sand dunes can be identified. Scientists can get a grasp of long term climate by studying sedimentary rock going back billions of years. The division of earth history into separate periods is largely based on visible changes in sedimentary rock layers that demarcate major changes in conditions. Often, they include major shifts in climate.

Sclerochronology

Corals

Coral "rings" are similar to tree rings except that they respond to different things, such as the water temperature, freshwater influx, pH changes, and wave action. From there, certain equipment can be used to derive the sea surface temperature and water salinity from the past few centuries. The $\delta^{18}O$ of coralline red algae provides a useful proxy of the combined sea surface temperature and sea surface salinity at high latitudes and the tropics, where many traditional techniques are limited.

Landscapes and Landforms

Within climatic geomorphology one approach is to study relict landforms to infer ancient climates. Being often concerned about past climates climatic geomorphology is considered sometimes to be a theme of historical geology. Climatic geomorphology is of limited use to study recent (Quaternary, Holocene) large climate changes since there are seldom discernible in the geomorphological record.

Time Scale and Limitations

A multinational consortium, the European Project for Ice Coring in Antarctica (EPICA), has drilled an ice core in Dome C on the East Antarctic ice sheet and retrieved ice from roughly 800,000 years ago. The international ice core community has, under the auspices of International Partnerships in Ice Core Sciences (IPICS), defined a priority project to obtain the oldest possible ice core record from Antarctica, an ice core record reaching back to or towards 1.5 million years ago. The deep marine record, the source of most isotopic data, exists only on oceanic plates, which are eventually subducted: the oldest remaining material is 200 million years old. Older sediments are also more prone to corruption by diagenesis. Resolution and confidence in the data decrease over time.

Notable Climate Events in Earth History

Knowledge of precise climatic events decreases as the record goes back in time, but some notable climate events are known:

- Faint young Sun paradox (start).
- Huronian glaciation (~2400 Mya Earth completely covered in ice probably due to Great Oxygenation Event).

- Later Neoproterozoic Snowball Earth (~600 Mya, precursor to the Cambrian Explosion).
- Andean-Saharan glaciation (~450 Mya).
- Carboniferous Rainforest Collapse (~300 Mya).
- Permian–Triassic extinction event (251.4 Mya).
- Oceanic anoxic events (~120 Mya, 93 Mya, and others).
- Cretaceous–Paleogene extinction event (66 Mya).
- Paleocene–Eocene Thermal Maximum (Paleocene–Eocene, 55Mya).
- Younger Dryas/The Big Freeze (~11,000 BC).
- Holocene climatic optimum (~7000–3000 BC).
- Extreme weather events of 535–536 (535–536 AD).
- Medieval Warm Period (900–1300).
- Little Ice Age (1300–1800).
- Year Without a Summer (1816).

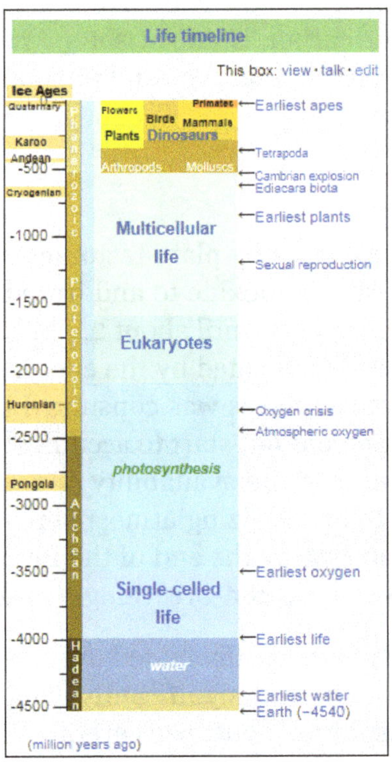

History of the atmosphere.

Earliest Atmosphere

The first atmosphere would have consisted of gases in the solar nebula, primarily hydrogen. In addition, there would probably have been simple hydrides such as those now found in gas giants like

Jupiter and Saturn, notably water vapor, methane, and ammonia. As the solar nebula dissipated, the gases would have escaped, partly driven off by the solar wind.

Second Atmosphere

The next atmosphere, consisting largely of nitrogen, carbon dioxide, and inert gases, was produced by outgassing from volcanism, supplemented by gases produced during the late heavy bombardment of Earth by huge asteroids. A major part of carbon dioxide emissions were soon dissolved in water and built up carbonate sediments.

Water-related sediments have been found dating from as early as 3.8 billion years ago. About 3.4 billion years ago, nitrogen was the major part of the then stable "second atmosphere". An influence of life has to be taken into account rather soon in the history of the atmosphere because hints of early life forms have been dated to as early as 3.5 billion years ago. The fact that it is not perfectly in line with the 30% lower solar radiance (compared to today) of the early Sun has been described as the "faint young Sun paradox".

The geological record, however, shows a continually relatively warm surface during the complete early temperature record of Earth with the exception of one cold glacial phase about 2.4 billion years ago. In the late Archaean eon, an oxygen-containing atmosphere began to develop, apparently from photosynthesizing cyanobacteria which have been found as stromatolite fossils from 2.7 billion years ago. The early basic carbon isotopy (isotope ratio proportions) was very much in line with what is found today, suggesting that the fundamental features of the carbon cycle were established as early as 4 billion years ago.

Third Atmosphere

The constant rearrangement of continents by plate tectonics influences the long-term evolution of the atmosphere by transferring carbon dioxide to and from large continental carbonate stores. Free oxygen did not exist in the atmosphere until about 2.4 billion years ago, during the Great Oxygenation Event, and its appearance is indicated by the end of the banded iron formations. Until then, any oxygen produced by photosynthesis was consumed by oxidation of reduced materials, notably iron. Molecules of free oxygen did not start to accumulate in the atmosphere until the rate of production of oxygen began to exceed the availability of reducing materials. That point was a shift from a reducing atmosphere to an oxidizing atmosphere. O2 showed major variations until reaching a steady state of more than 15% by the end of the Precambrian. The following time span was the Phanerozoic eon, during which oxygen-breathing metazoan life forms began to appear.

The amount of oxygen in the atmosphere has fluctuated over the last 600 million years, reaching a peak of 35% during the Carboniferous period, significantly higher than today's 21%. Two main processes govern changes in the atmosphere: plants use carbon dioxide from the atmosphere, releasing oxygen and the breakdown of pyrite and volcanic eruptions release sulfur into the atmosphere, which oxidizes and hence reduces the amount of oxygen in the atmosphere. However, volcanic eruptions also release carbon dioxide, which plants can convert to oxygen. The exact cause of the variation of the amount of oxygen in the atmosphere is not known. Periods with much oxygen in the atmosphere are associated with rapid development of animals. Today's atmosphere contains 21% oxygen, which is high enough for rapid development of animals.

Climate During Geological Ages

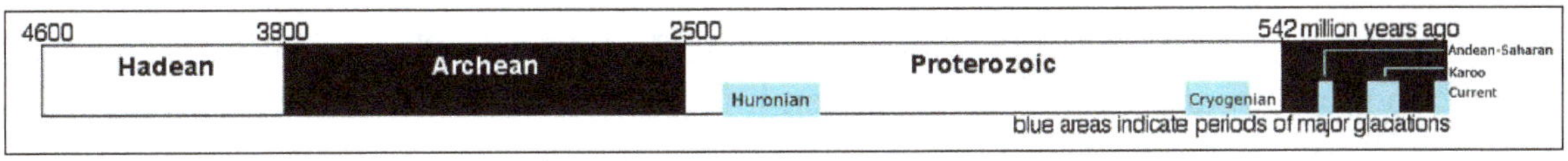

Timeline of glaciations, shown in blue.

- The Huronian glaciation, is the first known glaciation in Earth's history, and lasted from 2400-2100 million years ago.
- The Cryogenian glaciation lasted from 720-635 million years ago.
- The Andean-Saharan glaciation lasted from 450–420 million years ago.
- The Karoo glaciation lasted from 360–260 million years ago.
- The Quaternary glaciation is the current glaciation period and began 2.58 million years ago.

Precambrian Climate

The climate of the late Precambrian showed some major glaciation events spreading over much of the earth. At this time the continents were bunched up in the Rodinia supercontinent. Massive deposits of tillites and anomalous isotopic signatures are found, which gave rise to the Snowball Earth hypothesis. As the Proterozoic Eon drew to a close, the Earth started to warm up. By the dawn of the Cambrian and the Phanerozoic, life forms were abundant in the Cambrian explosion with average global temperatures of about 22 °C.

Phanerozoic Climate

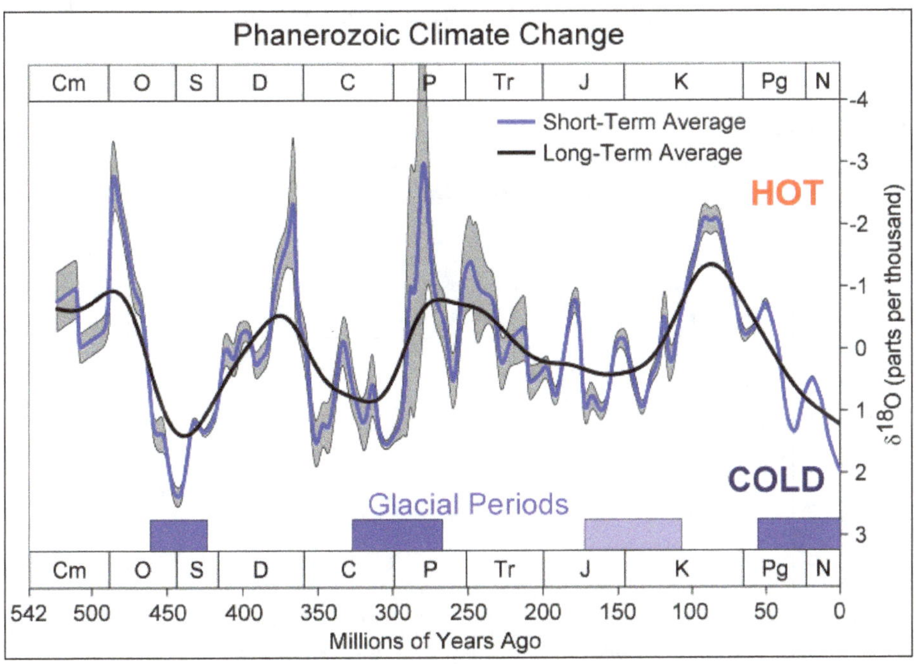

500 million years of climate change.

Major drivers for the preindustrial ages have been variations of the sun, volcanic ashes and exhalations, relative movements of the earth towards the sun, and tectonically induced effects as for major sea currents, watersheds, and ocean oscillations. In the early Phanerozoic, increased atmospheric carbon dioxide concentrations have been linked to driving or amplifying increased global temperatures. Royer et al. 2004 found a climate sensitivity for the rest of the Phanerozoic which was calculated to be similar to today's modern range of values.

The difference in global mean temperatures between a fully glacial Earth and an ice free Earth is estimated at approximately 10 °C, though far larger changes would be observed at high latitudes and smaller ones at low latitudes.One requirement for the development of large scale ice sheets seems to be the arrangement of continental land masses at or near the poles. The constant rearrangement of continents by plate tectonics can also shape long-term climate evolution. However, the presence or absence of land masses at the poles is not sufficient to guarantee glaciations or excludc polar ice caps. Evidence exists of past warm periods in Earth's climate when polar land masses similar to Antarctica were home to deciduous forests rather than ice sheets.

The relatively warm local minimum between Jurassic and Cretaceous goes along with an increase of subduction and mid-ocean ridge volcanism due to the breakup of the Pangea supercontinent.

Superimposed on the long-term evolution between hot and cold climates have been many short-term fluctuations in climate similar to, and sometimes more severe than, the varying glacial and interglacial states of the present ice age. Some of the most severe fluctuations, such as the Paleocene-Eocene Thermal Maximum, may be related to rapid climate changes due to sudden collapses of natural methane clathrate reservoirs in the oceans.

A similar, single event of induced severe climate change after a meteorite impact has been proposed as reason for the Cretaceous–Paleogene extinction event. Other major thresholds are the Permian-Triassic, and Ordovician-Silurian extinction events with various reasons suggested.

Quaternary Climate

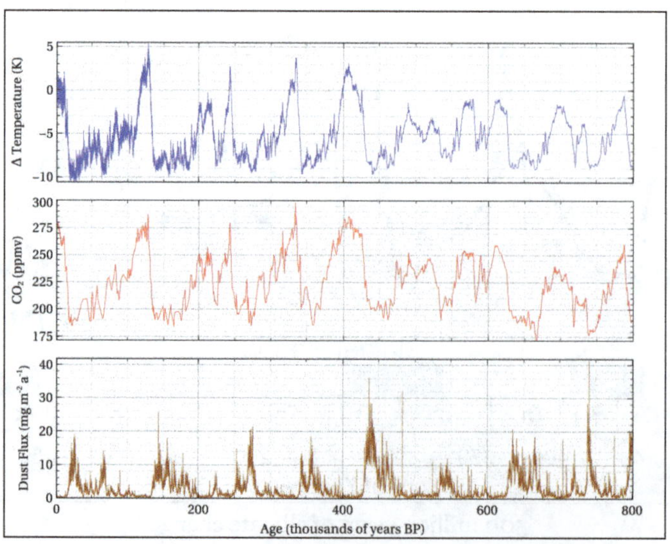

Ice core data for the past 800,000 years (x-axis values represent "age before 1950", so today's date is on the left side of the graph and older time on the right). Blue curve is temperature, red curve is atmospheric CO_2 concentrations, and brown curve is dust fluxes. Note length of glacial-interglacial cycles averages ~100,000 years.

The Quaternary sub-era includes the current climate. There has been a cycle of ice ages for the past 2.2–2.1 million years (starting before the Quaternary in the late Neogene Period).

Note in the graphic on the right the strong 120,000-year periodicity of the cycles, and the striking asymmetry of the curves. This asymmetry is believed to result from complex interactions of feedback mechanisms. It has been observed that ice ages deepen by progressive steps, but the recovery to interglacial conditions occurs in one big step.

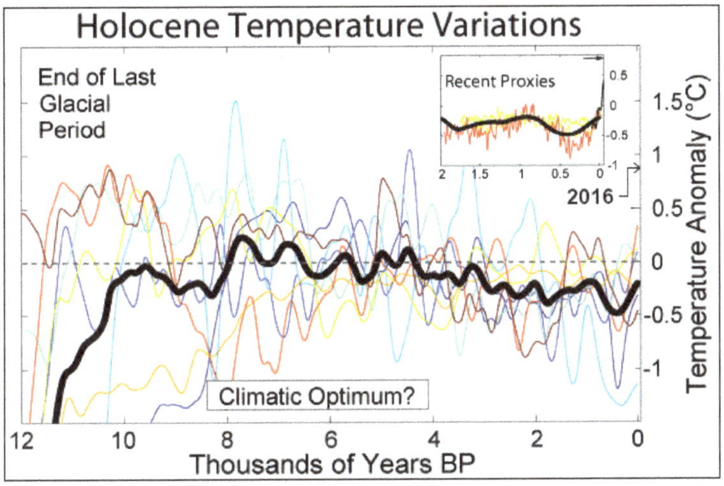

Holocene Temperature Variations.

The graph on the left shows the temperature change over the past 12,000 years, from various sources. The thick black curve is an average.

Climate Forcings

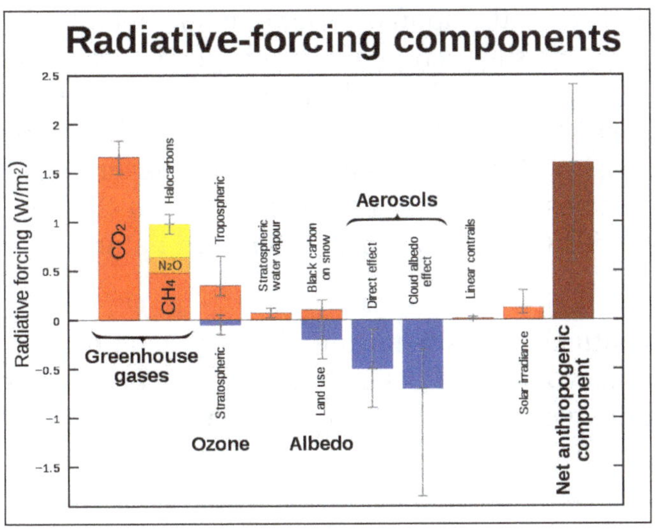

Radiative forcings, IPCC.

Climate forcing is the difference between radiant energy (sunlight) received by the Earth and the outgoing longwave radiation back to space. Radiative forcing is quantified based on the CO2 amount in the tropopause, in units of watts per square meter to the Earth's surface. Dependent on the radiative balance of incoming and outgoing energy, the Earth either warms up or cools down. Earth radiative balance originates from changes in solar insolation and the concentrations of greenhouse gases and aerosols. Climate change may be due to internal processes in Earth sphere's and/or following external forcings.

Internal Processes and Forcings

The Earth's climate system involves the atmosphere, biosphere, cryosphere, hydrosphere, and lithosphere, and the sum of these processes from Earth's spheres is what affects the climate. Greenhouse gasses act as the internal forcing of the climate system. Particular interests in climate science and paleoclimatology focus on the study of Earth climate sensitivity, in response to the sum of forcings.

Examples:

- Thermohaline circulation (Hydrosphere).
- Life (Biosphere).

External Forcings

- The Milankovitch cycles determine Earth distance and position to the Sun. The solar insolation is the total amount of solar radiation received by Earth.
- Volcanic eruptions are considered an external forcing.
- Human changes of the composition of the atmosphere or land use.

Mechanisms

On timescales of millions of years, the uplift of mountain ranges and subsequent weathering processes of rocks and soils and the subduction of tectonic plates, are an important part of the carbon cycle. The weathering sequesters CO_2, by the reaction of minerals with chemicals (especially silicate weathering with CO_2) and thereby removing CO_2 from the atmosphere and reducing the radiative forcing. The opposite effect is volcanism, responsible for the natural greenhouse effect, by emitting CO_2 into the atmosphere, thus affecting glaciation (Ice Age) cycles. James Hansen suggested that humans emit CO_2 10,000 times faster than natural processes have done in the past.

Ice sheet dynamics and continental positions (and linked vegetation changes) have been important factors in the long term evolution of the earth's climate. There is also a close correlation between CO_2 and temperature, where CO_2 has a strong control over global temperatures in Earth history.

ATMOSPHERIC CHEMISTRY

Atmospheric chemistry involves study of the chemistry of the atmospheres of Earth and other planets. It is a branch of atmospheric science and is a multidisciplinary field of research, drawing on environmental chemistry, meteorology, physics, computer modeling, oceanoraphy, geology, volcanology, and other disciplines. In addition, it is being increasingly associated with the field known as climatology.

Earth's atmosphere is composed of about 78 percent nitrogen, 21 percent oxygen, and small amounts of water vapor, carbon dioxide, argon, and other gases. This mixture of gases, commonly called air, protects and sustains life on Earth in a variety of ways. It provides oxygen for respiration, carbon dioxide for photosynthesis, and water vapor for the precipitation that replenishes moisture in the soil. In addition, carbon dioxide and water vapor act as "greenhouse gases" that keep the Earth sufficiently warm to maintain life. Nitrogen is used by "nitrogen-fixing" bacteria to produce compounds that are useful for plant growth. Water vapor prevents exposed living tissue from drying up. Ozone in the stratosphere absorbs ultraviolet solar radiation that could damage living tissue. In addition, higher layers of the atmosphere protect the Earth from bombardment by meteorites and charged particles in the solar wind.

The composition of Earth's atmosphere has been altered by human activities such as fuel burning and industrial production, and a number of these changes are harmful to human health, crops, and ecosystems. Examples of problems that involve studies in atmospheric chemistry include acid rain, photochemical smog, and global warming. Researchers in the field of atmospheric chemistry seek to understand the causes of these problems and to look for possible solutions. They help inform and evaluate government policies that are related to the environment.

Atmospheric Composition

Average composition of dry atmosphere (by volume)	
Gas	per NASA
Nitrogen, N_2	78.084%
Oxygen, O_2	20.946%
Argon, Ar	0.934%
Water vapor, H_2O	Highly variable; typically makes up about 1%
Minor constituents (in ppmv).	
Carbon Dioxide, CO_2	383
Neon, Ne	18.18
Helium, He	5.24
Methane, CH_4	1.7
Krypton, Kr	1.14
Hydrogen, H_2	0.55

- The concentration of CO_2 and CH_4 vary by season and location.
- ppmv represents parts per million by volume.
- The mean molecular mass of air is 28.97 g/mol.

Methodology

Observations, laboratory measurements, and modeling are the three central elements of atmospheric chemistry. Progress in this field is often driven by interactions between these components and they form an integrated whole. For example, observations may tell us that more of a chemical compound exists than previously thought possible. This would stimulate new modeling and laboratory studies, which would increase our scientific understanding to a point where the observations can be explained.

Observations

Observations are essential to our understanding of atmospheric chemistry. Routine observations of chemical composition provide information about changes in atmospheric composition over time. One important example of this is the Keeling Curve—a series of measurements from 1958 to today—that show a steady rise in the concentration of carbon dioxide.

These types of observations are conducted in observatories, such as that on Mauna Loa, and on mobile platforms such as aircraft (for instance, the UK's Facility for Airborne Atmospheric Measurements), ships, and balloons. Observations of atmospheric composition are increasingly made by satellites with important instruments, such as GOME and MOPITT, giving a global picture of air pollution and chemistry. Surface observations provide long-term records at high resolution in terms of time, but they are limited in the vertical and horizontal space they provide observations from. Some surface-based instruments, such as LIDAR, can provide concentration profiles of chemical compounds and aerosols, but they are restricted in the horizontal region they can cover. Many observations are available online in Atmospheric Chemistry Observational Databases.

Laboratory Measurements

Measurements made in the laboratory are essential to our understanding of the sources and sinks of pollutants and naturally occurring compounds. Lab studies tell us which gases react with one another and how fast they react. Measurements of interest include reactions in the gas phase, on surfaces, and in water. Of additional significance is photochemistry, which quantifies how quickly molecules are split apart by sunlight and the types of products formed, plus thermodynamic data such as Henry's law coefficients.

Modeling

To synthesize and test the theoretical understanding of atmospheric chemistry, computer models are constructed. Numerical models solve the differential equations governing the concentrations of chemicals in the atmosphere. They can range from simple to highly complex.

One common trade-off in numerical models is between the number of chemical compounds and chemical reactions modeled versus the representation of transport and mixing in the atmosphere. For example, a box model might include hundreds or even thousands of chemical reactions but will only have a very crude representation of mixing in the atmosphere. By contrast, 3D models represent many of the physical processes of the atmosphere but due to constraints on computer resources will have far fewer chemical reactions and compounds.

Models can be used to interpret observations, test understanding of chemical reactions, and predict future concentrations of chemical compounds in the atmosphere. One important current trend is for atmospheric chemistry modules to become one part of Earth system models in which the links between climate, atmospheric composition, and the biosphere can be studied.

Some models are constructed by automatic code generators. In this approach, a set of constituents are chosen and the automatic code generator then selects the reactions involving those constituents from a set of reaction databases. Once the reactions have been chosen, the ordinary differential equations (ODE) that describe the changes over time can be automatically constructed.

ATMOSPHERIC PHYSICS

Atmospheric physics is the application of physics to the study of the atmosphere. Atmospheric physicists attempt to model Earth's atmosphere and the atmospheres of the other planets using fluid flow equations, chemical models, radiation budget, and energy transfer processes in the atmosphere (as well as how these tie into other systems such as the oceans). In order to model weather systems, atmospheric physicists employ elements of scattering theory, wave propagation models, cloud physics, statistical mechanics and spatial statistics which are highly mathematical and related to physics. It has close links to meteorology and climatology and also covers the design and construction of instruments for studying the atmosphere and the interpretation of the data they provide, including remote sensing instruments. At the dawn of the space age and the introduction of sounding rockets, aeronomy became a subdiscipline concerning the upper layers of the atmosphere, where dissociation and ionization are important.

Remote Sensing

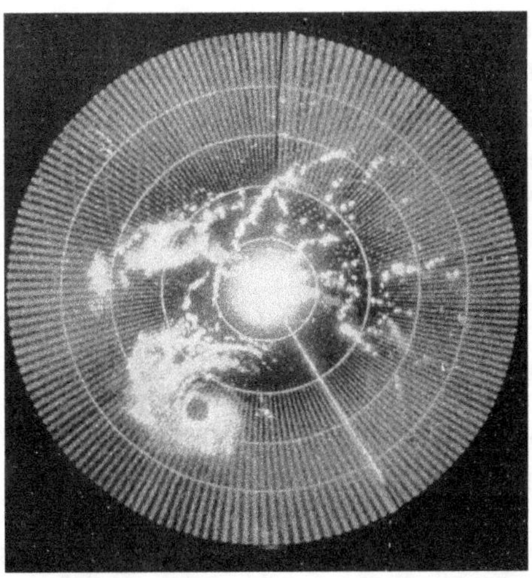

Brightness can indicate reflectivity as in this 1960 weather radar image (of Hurricane Abby). The radar's frequency, pulse form, and antenna largely determine what it can observe.

Remote sensing is the small or large-scale acquisition of information of an object or phenomenon, by the use of either recording or real-time sensing device(s) that is not in physical or intimate contact with the object (such as by way of aircraft, spacecraft, satellite, buoy, or ship). In practice, remote sensing is the stand-off collection through the use of a variety of devices for gathering information on a given object or area which gives more information than sensors at individual sites might convey. Thus, Earth observation or weather satellite collection platforms, ocean and atmospheric observing weather buoy platforms, monitoring of a pregnancy via ultrasound, magnetic resonance imaging (MRI), positron-emission tomography (PET), and space probes are all examples of remote sensing. In modern usage, the term generally refers to the use of imaging sensor technologies including but not limited to the use of instruments aboard aircraft and spacecraft, and is distinct from other imaging-related fields such as medical imaging.

There are two kinds of remote sensing. Passive sensors detect natural radiation that is emitted or reflected by the object or surrounding area being observed. Reflected sunlight is the most common source of radiation measured by passive sensors. Examples of passive remote sensors include film photography, infra-red, charge-coupled devices, and radiometers. Active collection, on the other hand, emits energy in order to scan objects and areas whereupon a sensor then detects and measures the radiation that is reflected or backscattered from the target. radar, lidar, and SODAR are examples of active remote sensing techniques used in atmospheric physics where the time delay between emission and return is measured, establishing the location, height, speed and direction of an object.

Remote sensing makes it possible to collect data on dangerous or inaccessible areas. Remote sensing applications include monitoring deforestation in areas such as the Amazon Basin, the effects of climate change on glaciers and Arctic and Antarctic regions, and depth sounding of coastal and ocean depths. Military collection during the Cold War made use of stand-off collection of data about dangerous border areas. Remote sensing also replaces costly and slow data collection on the ground, ensuring in the process that areas or objects are not disturbed.

Orbital platforms collect and transmit data from different parts of the electromagnetic spectrum, which in conjunction with larger scale aerial or ground-based sensing and analysis, provides researchers with enough information to monitor trends such as El Niño and other natural long and short term phenomena. Other uses include different areas of the earth sciences such as natural resource management, agricultural fields such as land usage and conservation, and national security and overhead, ground-based and stand-off collection on border areas.

Radiation

his is a diagram of the seasons. In addition to the density of incident light, the dissipation of light in the atmosphere is greater when it falls at a shallow angle.

Atmospheric physicists typically divide radiation into solar radiation (emitted by the sun) and terrestrial radiation (emitted by Earth's surface and atmosphere).

Solar radiation contains variety of wavelengths. Visible light has wavelengths between 0.4 and 0.7 micrometers. Shorter wavelengths are known as the ultraviolet (UV) part of the spectrum, while longer wavelengths are grouped into the infrared portion of the spectrum. Ozone is most effective in absorbing radiation around 0.25 micrometers, where UV-c rays lie in the spectrum. This increases the temperature of the nearby stratosphere. Snow reflects 88% of UV rays, while sand reflects 12%, and water reflects only 4% of incoming UV radiation. The more glancing the angle is between the atmosphere and the sun's rays, the more likely that energy will be reflected or absorbed by the atmosphere.

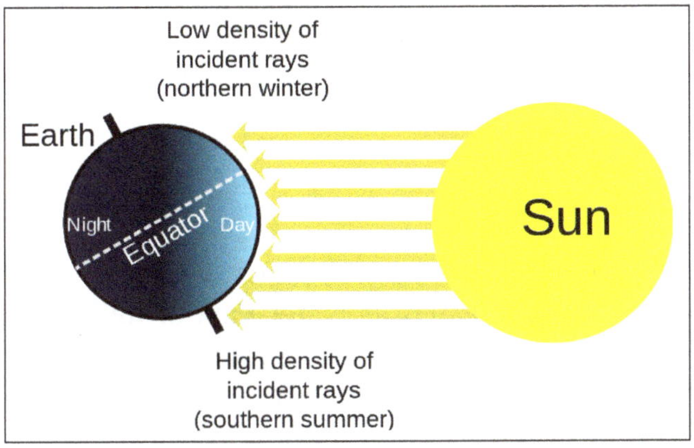

Terrestrial radiation is emitted at much longer wavelengths than solar radiation. This is because Earth is much colder than the sun. Radiation is emitted by Earth across a range of wavelengths, as formalized in Planck's law. The wavelength of maximum energy is around 10 micrometers.

Cloud Physics

Cloud physics is the study of the physical processes that lead to the formation, growth and precipitation of clouds. Clouds are composed of microscopic droplets of water (warm clouds), tiny crystals of ice, or both (mixed phase clouds). Under suitable conditions, the droplets combine to form precipitation, where they may fall to the earth. The precise mechanics of how a cloud forms and grows is not completely understood, but scientists have developed theories explaining the structure of clouds by studying the microphysics of individual droplets. Advances in radar and satellite technology have also allowed the precise study of clouds on a large scale.

Atmospheric Electricity

Cloud-to-ground lightning in the global atmospheric electrical circuit.

Atmospheric electricity is the term given to the electrostatics and electrodynamics of the atmosphere (or, more broadly, the atmosphere of any planet). The Earth's surface, the ionosphere, and the atmosphere is known as the global atmospheric electrical circuit. Lightning discharges 30,000 amperes, at up to 100 million volts, and emits light, radio waves, X-rays and even gamma rays. Plasma temperatures in lightning can approach 28,000 kelvins and electron densities may exceed $10^{24}/m^3$.

Atmospheric Tide

The largest-amplitude atmospheric tides are mostly generated in the troposphere and stratosphere when the atmosphere is periodically heated as water vapour and ozone absorb solar radiation during the day. The tides generated are then able to propagate away from these source regions and ascend into the mesosphere and thermosphere. Atmospheric tides can be measured as regular fluctuations in wind, temperature, density and pressure. Although atmospheric tides share much in common with ocean tides they have two key distinguishing features:

- Atmospheric tides are primarily excited by the Sun's heating of the atmosphere whereas ocean tides are primarily excited by the Moon's gravitational field. This means that most atmospheric tides have periods of oscillation related to the 24-hour length of the solar day whereas ocean tides have longer periods of oscillation related to the lunar day (time between successive lunar transits) of about 24 hours 51 minutes.
- Atmospheric tides propagate in an atmosphere where density varies significantly with height. A consequence of this is that their amplitudes naturally increase exponentially as the tide ascends into progressively more rarefied regions of the atmosphere. In contrast, the density of the oceans varies only slightly with depth and so there the tides do not necessarily vary in amplitude with depth.

Note that although solar heating is responsible for the largest-amplitude atmospheric tides, the gravitational fields of the Sun and Moon also raise tides in the atmosphere, with the lunar gravitational atmospheric tidal effect being significantly greater than its solar counterpart.

At ground level, atmospheric tides can be detected as regular but small oscillations in surface pressure with periods of 24 and 12 hours. Daily pressure maxima occur at 10 a.m. and 10 p.m. local time, while minima occur at 4 a.m. and 4 p.m. local time. The absolute maximum occurs at 10 a.m. while the absolute minimum occurs at 4 p.m. However, at greater heights the amplitudes of the tides can become very large. In the mesosphere (heights of ~ 50 – 100 km) atmospheric tides can reach amplitudes of more than 50 m/s and are often the most significant part of the motion of the atmosphere.

Aeronomy

Aeronomy is the science of the upper region of the atmosphere, where dissociation and ionization are important. The term aeronomy was introduced by Sydney Chapman in 1960. Today, the term also includes the science of the corresponding regions of the atmospheres of other planets. Research in aeronomy requires access to balloons, satellites, and sounding rockets which provide valuable data about this region of the atmosphere. Atmospheric tides play an important role in interacting with both the lower and upper atmosphere. Amongst the phenomena studied are upper-atmospheric lightning discharges, such as luminous events called red sprites, sprite halos, blue jets, and elves.

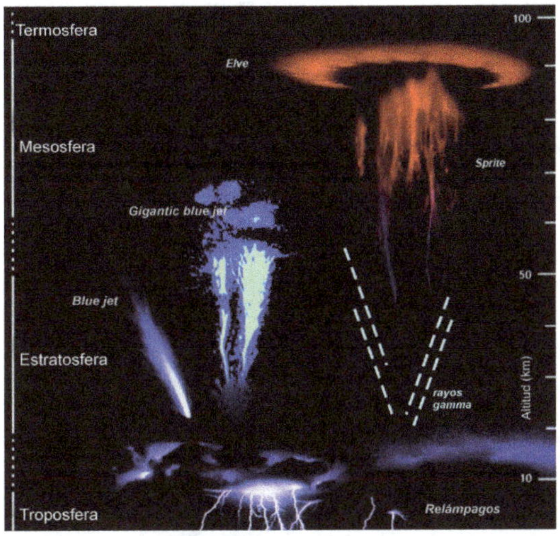

Representation of upper-atmospheric lightning and electrical-discharge phenomena.

ATMOSPHERIC DYNAMICS

Atmospheric dynamics involves observational and theoretical analysis of all motion systems of meteorological significance, including such diverse phenomena as thunderstorms, tornadoes, gravity waves,tropical hurricanes, extratropical cyclones, jet streams, and global-scale circulations. The immediate goal of dynamical studies is to explain the observed circulations on the basis of fundamental physical principles. The practical objectives of such studies include improving weather prediction, developing methods for prediction of short-term (seasonal and interannual) climate fluctuations, and understanding the implications of human-induced perturbations (e.g., increased carbon dioxide concentrations or depletion of the ozone layer) on the global climate.

CONIOLOGY

Coniology or koniology is the study of atmospheric dust and its effects. Samples of dust are often collected by a device called a coniometer. Coniology refers to the observation and contemplation of dust in an atmosphere, but the study of dust may also be applied to dust in space, therefore connecting it to a variety of atmospheric and extraterrestrial topics.

Earth

Dust in Earth's atmosphere comes from both natural and anthropogenic causes. The process of which dust enters Earth's atmosphere naturally can be attributed to the Aeolian Process where winds erode Earth's surface and consequently carry particles from the ground into the atmosphere via suspension. Where as anthropogenic dust may be due to human activities such as farming and creating air pollutants. The solid dust particulates and atmospheric moisture combined are suspended in the air which are both natural and anthropogenic are referred to as Aerosols. The Sahara Desert is one of the biggest contributors in the creation of atmospheric dust as the winds carry and deposit dust and particulates across the planet.

A dust storm traveling from the Sahara Desert over the Atlantic Ocean.

The effect dust has on living organisms is both observed positively and negatively. The dust that is carried by wind from the Sahara desert can travel as far as the Caribbean and South America. This dust carries nutrients which further supports the ecosystems in the Caribbean, and those on the South American continent. Furthermore, nutrients in dust can carried by winds which reach all across the globe. On the other hand, dust can also be more harmful for an ecosystem depending on what it carries. Research conducted by the senior geologist at the U.S. Geological Survey Center for Coastal Geology, Gene Shinn, revealed a correlation of decline in coral reefs with increased harmful pathogens to coral transported by dust from the Sahara to the Caribbean, coupled with an increasing global ocean temperature.

Atmospheric dust, both natural and anthropogenic, has also been shown to effect humans as well. One of the biggest contributing factors that the effect atmospheric dust has on humans is health, more specifically respiratory health. Barbados and Trinidad have seen a rise in asthma in their population and one of the speculated causes is an influx of dust being carried over the Atlantic from the Sahara desert. Certain Particulates, or particulate matter PM have lead to health issues among humans such as lung disease and cardiovascular disease.

Moon

The study of dust and its effect is not only applied to the Earth, but also the Moon. One NASA mission sent a spacecraft to study the moon's atmosphere, this spacecraft was called LADEE. LADEE's mission revealed to scientist how different the moon's atmosphere is compared to the Earth, which may give insight on other celestial bodies and their atmospheric composition. The Moons atmosphere is very tenuous and creates an atmosphere with minimal dust presence. The moon can credit its thin atmosphere to solar waves that deliver hydrogen and helium to its exosphere and gasses of argon and helium being released from the lunar rocks. Even more interesting, is the fluctuation of the moons atmospheric content in relation to the dust and elements in a given time. Of the most abundant elements in the moon's atmosphere, helium, argon and neon, will all be present but at different amounts in the atmosphere depending on where and how long the suns ray hit the moon.

Other Celestial Bodies

Space is known for its vast emptiness; however, there is an enormous amount of dust in space, whether it be on comets, moons, planets, or nebulae. The Herschel Space Observatory provided scientist with data about how celestial bodies formed in space, while also making new discoveries pertaining to dust in space. Dust in space, or cosmic dust create mass required for the formation of all celestial bodies like planets and stars. From the research gathered by the Herschel Space Observatory, Supernovae, or the massive explosion of a star that occurs at the end of a stars life, was the biggest contributing factor of ejecting dust and elements into space essentially creating the make up for matter in our universe.

A section of the Carina Nebula.

When attempting to study space, cosmic dust would tend to hinder the scientists attempting to observe the celestial bodies in space. However, cosmic dust went from a hindrance, to a new and important aspect of scientific study. The Herschel Space Observatory used far reaching infrared equipment which revealed light being emitted from cosmic dust and gasses. This infrared technology allowed the Herschel Space Observatory and scientist to see more objects in space that emit heat at lower temperatures, which in turn allows more celestial bodies to be seen when they would not have been without the advanced infrared technologies.

Another spacecraft that helped further our understanding of cosmic dust was the Stardust spacecraft. Stardust's mission was to collect samples of cosmic dust from the Wild-2 comet. The reason why this mission is significant is it will not only help scientist to further understand the properties of dust in space, but it was the first mission to collect dust from a comet and return it to Earth.

References

- Meteorology, encyclopedia: nationalgeographic.org, Retrieved 25 August, 2019
- Bradley, raymond (2015). Paleoclimatology: reconstructing climates of the quaternary. Oxford: elsevier. P. 1. Isbn 978-0-12-386913-5

- Climatology: environmentalscience.org, Retrieved 4 February, 2019
- Beerling, david (2007). The emerald planet: how plants changed earth's history. Oxford university press. P. 47. Isbn 9780192806024
- Zahnle, k.; schaefer, l.; fegley, b. (2010). "earth's earliest atmospheres". Cold spring harbor perspectives in biology. 2 (10): a004895. Doi:10.1101/cshperspect.a004895. Pmc 2944365. Pmid 20573713
- Atmospheric-chemistry, entry: newworldencyclopedia.org, Retrieved 12 May, 2019
- Wheeling jesuit university. Exploring the environment: uv menace.archived august 30, 2007, at the wayback machine retrieved on 2007-06-01
- Atmosdyn, academic: washington.edu, Retrieved 1 May, 2019
- Culler, jessica (2015-06-16). "ladee - lunar atmosphere dust and environment explorer". Nasa. Retrieved 2019-04-14

3

Layers of Earth's Atmosphere

The atmosphere of the Earth is divided into various layers, all of which have their own specific traits. Some of these layers are the troposphere, stratosphere, mesosphere, thermosphere, exosphere, ozonosphere and magnetosphere. This chapter has been carefully written to provide an easy understanding of these layers of the Earth's atmosphere.

Earth's atmosphere has a series of layers, each with its own specific traits. Moving upward from ground level, these layers are named the troposphere, stratosphere, mesosphere, thermosphere and exosphere. The exosphere gradually fades away into the realm of interplanetary space.

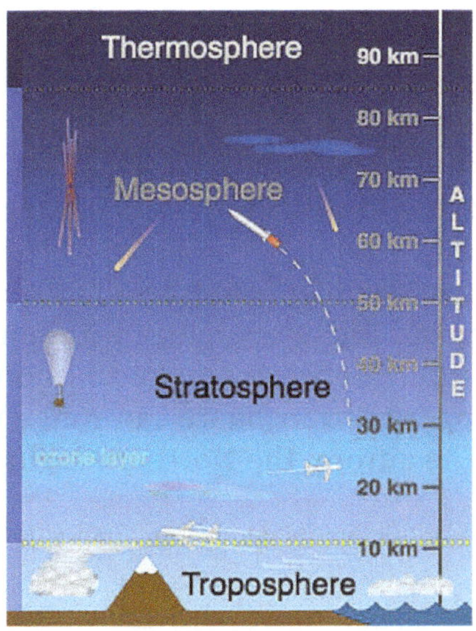

Layers of Earth's Atmosphere Layers of the atmosphere: troposphere, stratosphere, mesosphere and thermosphere.

Troposphere

The troposphere is the lowest layer of our atmosphere. Starting at ground level, it extends upward to about 10 km (6.2 miles or about 33,000 feet) above sea level. We humans live in the troposphere, and nearly all weather occurs in this lowest layer. Most clouds appear here, mainly because 99% of the water vapor in the atmosphere is found in the troposphere. Air pressure drops, and temperatures get colder, as you climb higher in the troposphere.

Stratosphere

The next layer up is called the stratosphere. The stratosphere extends from the top of the troposphere to about 50 km (31 miles) above the ground. The infamous ozone layer is found within the stratosphere. Ozone molecules in this layer absorb high-energy ultraviolet (UV) light from the Sun, converting the UV energy into heat. Unlike the troposphere, the stratosphere actually gets warmer the higher you go! That trend of rising temperatures with altitude means that air in the stratosphere lacks the turbulence and updrafts of the troposphere beneath. Commercial passenger jets fly in the lower stratosphere, partly because this less-turbulent layer provides a smoother ride. The jet stream flows near the border between the troposphere and the stratosphere.

Mesosphere

Above the stratosphere is the mesosphere. It extends upward to a height of about 85 km (53 miles) above our planet. Most meteors burn up in the mesosphere. Unlike the stratosphere, temperatures once again grow colder as you rise up through the mesosphere. The coldest temperatures in Earth's atmosphere, about -90° C (-130° F), are found near the top of this layer. The air in the mesosphere is far too thin to breathe; air pressure at the bottom of the layer is well below 1% of the pressure at sea level, and continues dropping as you go higher.

Thermosphere

The layer of very rare air above the mesosphere is called the thermosphere. High-energy X-rays and UV radiation from the Sun are absorbed in the thermosphere, raising its temperature to hundreds or at times thousands of degrees. However, the air in this layer is so thin that it would feel freezing cold to us! In many ways, the thermosphere is more like outer space than a part of the atmosphere. Many satellites actually orbit Earth within the thermosphere! Variations in the amount of energy coming from the Sun exert a powerful influence on both the height of the top of this layer and the temperature within it. Because of this, the top of the thermosphere can be found anywhere between 500 and 1,000 km (311 to 621 miles) above the ground. Temperatures in the upper thermosphere can range from about 500° C (932° F) to 2,000° C (3,632° F) or higher. The aurora, the Northern Lights and Southern Lights, occur in the thermosphere.

Exosphere

Although some experts consider the thermosphere to be the uppermost layer of our atmosphere, others consider the exosphere to be the actual "final frontier" of Earth's gaseous envelope. As you might imagine, the "air" in the exosphere is very, very, very thin, making this layer even more space-like than the thermosphere. In fact, air in the exosphere is constantly - though very gradually - "leaking" out of Earth's atmosphere into outer space. There is no clear-cut upper boundary where the exosphere finally fades away into space. Different definitions place the top of the exosphere somewhere between 100,000 km (62,000 miles) and 190,000 km (120,000 miles) above the surface of Earth. The latter value is about halfway to the Moon.

Ionosphere

The ionosphere is not a distinct layer like the others mentioned above. Instead, the ionosphere is a series of regions in parts of the mesosphere and thermosphere where high-energy radiation from the Sun has knocked electrons loose from their parent atoms and molecules. The electrically charged atoms and molecules that are formed in this way are called ions, giving the ionosphere its name and endowing this region with some special properties.

TROPOSPHERE

The lowest layer of the atmosphere is called the troposphere. It ranges in thickness from 8km at the poles to 16km over the equator. The troposphere is bounded above by the tropopause, a boundary marked by stable temperatures. Above the troposphere is the stratosphere. Although variations do occur, temperature usually declines with increasing altitude in the troposphere. Hill walkers know that it will be several degrees cooler on the top of a mountain than in the valley below.

The troposphere is denser than the layers of the atmosphere above it (because of the weight compressing it), and it contains up to 75% of the mass of the atmosphere. It is primarily composed of nitrogen (78%) and oxygen (21%) with only small concentrations of other trace gases. Nearly all atmospheric water vapour or moisture is found in the troposphere.

The troposphere is the layer where most of the world's weather takes place. Since temperature decreases with altitude in the troposphere, warm air near the surface of the Earth can readily rise, being less dense than the colder air above it. In fact air molecules can travel to the top of the troposphere and back down again in a just a few days. Such vertical movement or convection of air generates clouds and ultimately rain from the moisture within the air, and gives rise to much of the weather which we experience. The troposphere is capped by the tropopause, a region of stable temperature. Air temperature then begins to rise in the stratosphere. Such a temperature increase prevents much air convection beyond the tropopause, and consequently most weather phenomena, including towering cumulonimbus thunderclouds, are confined to the troposphere.

Sometimes the temperature does not decrease with height in the troposphere, but increases. Such a situation is known as a temperature inversion. Temperature inversions limit or prevent the vertical mixing of air. Such atmospheric stability can lead to air pollution episodes with air pollutants emitted at ground level becoming trapped underneath the temperature inversion.

STRATOSPHERE

The stratosphere is a layer of Earth's atmosphere. It is the second layer of the atmosphere as you go upward. The troposphere, the lowest layer, is right below the stratosphere. The next higher layer above the stratosphere is the mesosphere.

The bottom of the stratosphere is around 10 km (6.2 miles or about 33,000 feet) above the ground at middle latitudes. The top of the stratosphere occurs at an altitude of 50 km (31 miles). The height of the bottom of the stratosphere varies with latitude and with the seasons. The lower boundary of the stratosphere can be as high as 20 km (12 miles or 65,000 feet) near the equator and as low as 7 km (4 miles or 23,000 feet) at the poles in winter. The lower boundary of the stratosphere is called the tropopause; the upper boundary is called the stratopause.

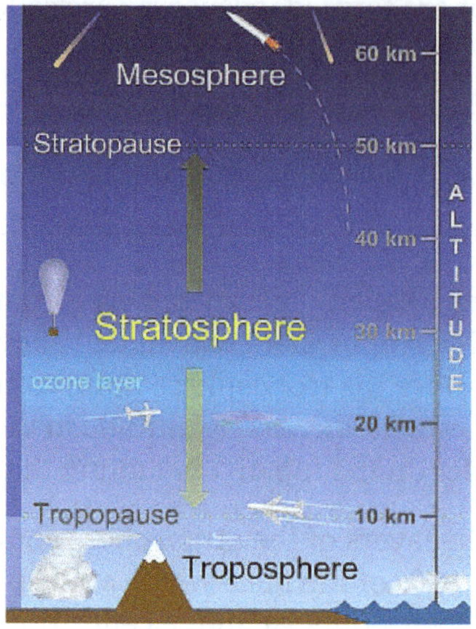

This diagram shows some of the features of the stratosphere.

Ozone, an unusual type of oxygen molecule that is relatively abundant in the stratosphere, heats this layer as it absorbs energy from incoming ultraviolet radiation from the Sun. Temperatures rise as one moves upward through the stratosphere. This is exactly the opposite of the behavior in the troposphere in which we live, where temperatures drop with increasing altitude. Because of this temperature stratification, there is little convection and mixing in the stratosphere, so the layers of air there are quite stable. Commercial jet aircraft fly in the lower stratosphere to avoid the turbulence which is common in the troposphere below.

The stratosphere is very dry; air there contains little water vapor. Because of this, few clouds are found in this layer; almost all clouds occur in the lower, more humid troposphere. Polar stratospheric clouds (PSCs) are the exception. PSCs appear in the lower stratosphere near the poles in winter. They are found at altitudes of 15 to 25 km (9.3 to 15.5 miles) and form only when temperatures at those heights dip below -78° C. They appear to help cause the formation of the infamous holes in the ozone layer by "encouraging" certain chemical reactions that destroy ozone. PSCs are also called nacreous clouds.

Air is roughly a thousand times thinner at the top of the stratosphere than it is at sea level. Because of this, jet aircraft and weather balloons reach their maximum operational altitudes within the stratosphere.

Due to the lack of vertical convection in the stratosphere, materials that get into the stratosphere can stay there for long times. Such is the case for the ozone-destroying chemicals called CFCs (chlorofluorocarbons). Large volcanic eruptions and major meteorite impacts can fling aerosol particles up into the stratosphere where they may linger for months or years, sometimes altering Earth's global climate. Rocket launches inject exhaust gases into the stratosphere, producing uncertain consequences.

Various types of waves and tides in the atmosphere influence the stratosphere. Some of these waves and tides carry energy from the troposphere upward into the stratosphere; others convey energy from the stratosphere up into the mesosphere. The waves and tides influence the flows of air in the stratosphere and can also cause regional heating of this layer of the atmosphere.

A rare type of electrical discharge, somewhat akin to lightning, occurs in the stratosphere. These "blue jets" appear above thunderstorms, and extend from the bottom of the stratosphere up to altitudes of 40 or 50 km (25 to 31 miles).

MESOSPHERE

The mesosphere is one of five layers of the atmosphere surrounding the planet earth. The other four layers include the troposphere, stratosphere, thermosphere, and exosphere. The mesosphere is located approximately 50km from the earth's surface and extends as far as 85km from the earth's surface. The mesosphere is located above the stratosphere and below the thermosphere. And it is third layer, located above the troposphere and stratosphere. The mesosphere is too high for weather balloons or airplanes and too low for satellites, making it difficult to study, however scientists are able to use research rockets capable of short trips for specific experiments.

Due to high atmospheric drag in the mesosphere it is not possible for research equipment such as satellites to stay in orbit.

The mesosphere is very important for earth's protection. The mesosphere burns up most meteors and asteroids before they are able to reach the earth's surface.

It is estimated that approximately 40 tons of meteors fall towards earth each day, and the mesosphere is responsible for burning them up before they reach the earth and cause damage to its surface.

As meteors burn up they can sometimes be seen in the night's sky. Most people call them shooting stars.

The mesosphere's atmosphere is low density and made up of oxygen, carbon dioxide, and nitrogen.

Although most of the meteors that reach the mesosphere are burned up, some of their material hangs around afterwards, including iron and other metallic atoms.

The temperature of the mesosphere becomes colder as the distance from the earth increases. The temperature can drop to -140 degrees Celsius however at its warmest level, depending on season, the temperature can reach -5 degrees Celsius.

The mesosphere is the coldest atmospheric layer surrounding the earth. It becomes cold enough to freeze water vapour in its atmosphere into ice clouds. These ice clouds are blue-white and are called noctilucent clouds or polar mesospheric clouds. These clouds are more visible at sunset from the earth's poles.

The mesosphere experiences atmospheric gravity waves, atmospheric tides, planetary waves, and strong winds that flow from north to south and east to west called zonal winds.

The research rockets used to study the mesosphere are also called sounding rockets. These rockets are often made with surplus military rocket motors.

A strange type of lightning occurs in the mesosphere. This lightning is referred to as 'sprites' or 'elves'.

Together the layers of the atmosphere help to protect the earth from greenhouse gases, working like a blanket of insulation surrounding the planet.

The atmosphere around the earth, including the mesosphere, helps to keep the earth's climate and weather patterns as regular as possible.

The area where the mesosphere transitions into the thermosphere is called the mesopause. This is the coldest area of the mesosphere.

In the lower mesosphere the zonal winds blow from the north to the south, while in the upper mesosphere they blow from east to west.

THERMOSPHERE

The aurora (Northern Lights and Southern Lights) mostly occur in the thermosphere.

The thermosphere is a layer of Earth's atmosphere. The thermosphere is directly above the mesosphere and below the exosphere. It extends from about 90 km (56 miles) to between 500 and 1,000 km (311 to 621 miles) above our planet.

Temperatures climb sharply in the lower thermosphere (below 200 to 300 km altitude), then level off and hold fairly steady with increasing altitude above that height. Solar activity strongly influences temperature in the thermosphere. The thermosphere is typically about 200° C (360° F) hotter in the daytime than at night, and roughly 500° C (900° F) hotter when the Sun is very active than at other times. Temperatures in the upper thermosphere can range from about 500° C (932° F) to 2,000° C (3,632° F) or higher.

The boundary between the thermosphere and the exosphere above it is called the thermopause. At the bottom of the thermosphere is the mesopause, the boundary between the thermosphere and the mesosphere below.

Although the thermosphere is considered part of Earth's atmosphere, the air density is so low in this layer that most of the thermosphere is what we normally think of as outer space. In fact, the most common definition says that space begins at an altitude of 100 km (62 miles), slightly above the mesopause at the bottom of the thermosphere. The space shuttle and the International Space Station both orbit Earth within the thermosphere.

Below the thermosphere, gases made of different types of atoms and molecules are thoroughly mixed together by turbulence in the atmosphere. Air in the lower atmosphere is mainly composed of the familiar blend of about 80% nitrogen molecules (N2) and about 20% oxygen molecules (O2). In the thermosphere and above, gas particles collide so infrequently that the gases become somewhat separated based on the types of chemical elements they contain. Energetic ultraviolet and X-ray photons from the Sun also break apart molecules in the thermosphere. In the upper thermosphere, atomic oxygen (O), atomic nitrogen (N), and helium (He) are the main components of air.

Much of the X-ray and UV radiation from the Sun is absorbed in the thermosphere. When the Sun is very active and emitting more high energy radiation, the thermosphere gets hotter and expands or "puffs up". Because of this, the height of the top of the thermosphere (the thermopause) varies. The thermopause is found at an altitude between 500 km and 1,000 km or higher. Since many satellites orbit within the thermosphere, changes in the density of (the very, very thin) air at orbital altitudes brought on by heating and expansion of the thermosphere generates a drag force on satellites. Engineers must take this varying drag into account when calculating orbits, and satellites occasionally need to be boosted higher to offset the effects of the drag force.

High-energy solar photons also tear electrons away from gas particles in the thermosphere, creating electrically-charged ions of atoms and molecules. Earth's ionosphere, composed of several regions of such ionized particles in the atmosphere, overlaps with and shares the same space with the electrically neutral thermosphere.

Like the oceans, Earth's atmosphere has waves and tides within it. These waves and tides help move energy around within the atmosphere, including the thermosphere. Winds and the overall circulation in the thermosphere are largely driven by these tides and waves. Moving ions, dragged along by collisions with the electrically neutral gases, produce powerful electrical currents in some parts of the thermosphere.

Finally, the aurora (the Southern and Northern Lights) primarily occur in the thermosphere. Charged particles (electrons, protons, and other ions) from space collide with atoms and molecules in the thermosphere at high latitudes, exciting them into higher energy states. Those atoms and molecules shed this excess energy by emitting photons of light, which we see as colorful auroral displays.

EXOSPHERE

The exosphere is the outermost layer of the Earth's atmosphere, located above the thermosphere. It extends from about 600 km until it thins out to merge with interplanetary space. This makes the exosphere about 10,000 km or 6,200 miles thick or about as wide as the Earth. The top boundary of Earth's exosphere extends about halfway to the Moon.

For other planets with substantial atmospheres, the exosphere is the layer above the denser atmospheric layers, but for planets or satellites without dense atmospheres, the exosphere is the region between the surface and interplanetary space. This is called the surface boundary exosphere. It has been observed for the Earth's Moon, Mercury, and the Galilean moons of Jupiter.

Exosphere Characteristics

The particles in the exosphere are extremely far apart. They don't quite fit the definition of a "gas" because the density is too low for collisions and interactions to occur. Nor are they necessarily plasma, because the atoms and molecules aren't all electrically charged. Particles in the exosphere can travel hundreds of kilometers along a ballistic trajectory before bumping into other particles.

The Earth's Exosphere

The lower boundary of the exosphere, where it meets the thermosphere, is called the thermopause. Its height above sea level ranges from 250-500 km up to 1000 km (310 to 620 miles), depending on solar activity. The thermopause is called the exobase, exopause, or critical altitude. Above this point, barometric conditions do not apply. The temperature of the exosphere is nearly constant and very cold. At the upper boundary of the exosphere, the solar radiation pressure on hydrogen exceeds the gravitational pull back toward Earth. The fluctuation of the exobase due to solar weather is important because it affects atmospheric drag on space stations and satellites. Particles that reach the boundary are lost from the Earth's atmosphere to space.

The composition of the exosphere is different from that of the layers beneath it. Only the lightest gases occur, barely held to the planet by gravity. The Earth's exosphere consists mainly of hydrogen, helium, carbon dioxide, and atomic oxygen. The exosphere is visible from space as a fuzzy region called the geocorona.

The Lunar Atmosphere

One Earth, there are about 10^{19} molecules per cubic centimeter of air at sea level. In contrast, there are fewer than a million (10^6) molecules in the same volume in the exosphere.The Moon does not have a true atmosphere because its particles don't circulate, don't absorb much radiation, and have to be replenished. Yet, it's not quite a vacuum, either. The lunar surface boundary layer has a pressure of about 3×10^{-15} atm (0.3 nano Pascals). The pressure varies depending on whether it's day or night, but the entire mass weighs less than 10 metric tonnes. The exosphere is produced by outgassing of radon and helium from radioactive decay. The solar wind, micrometeor bombardment, and the solar wind also contribute particles. Unusual gases found in the Moon's exosphere, but not in the atmosphere's of Earth, Venus, or Mars include sodium and potassium. Other elements and compounds found in the Moon's exosphere include argon-40, neon, helium-4, oxygen, methane, nitrogen, carbon monoxide, and carbon dioxide.

A trace amount of hydrogen is present. Very minute quantities of water vapor may also exist.

In addition to its exosphere, the Moon may have an "atmosphere" of dust that hovers above the surface due to electrostatic levitation.

IONOSPHERE AND MAGNETOSPHERE

Ionosphere and magnetosphere are the regions of Earth's atmosphere in which the number of electrically charged particles—ions and electrons—are large enough to affect the propagation of radio waves. The charged particles are created by the action of extraterrestrial radiation (mainly from the Sun) on neutral atoms and molecules of air. The ionosphere begins at a height of about 50 km (30 miles) above the surface, but it is most distinct and important above 80 km (50 miles). In the upper regions of the ionosphere, beginning several hundred kilometres above Earth's surface and extending tens of thousands of kilometres into space, is the magnetosphere, a region where the behaviour of charged particles is strongly affected by the magnetic fields of Earth and the Sun. It is in the lower part of the magnetosphere that overlaps with the ionosphere that the spectacular displays of the aurora borealis and aurora australis take place. The magnetosphere also contains the Van Allen radiation belts, where highly energized protons and electrons travel back and forth between the poles of Earth's magnetic field.

Ionosphere

Discovery of the Ionosphere

Discovery of the ionosphere extended over nearly a century. As early as 1839, the German mathematician Carl Friedrich Gauss speculated that an electrically conducting region of the atmosphere could account for observed variations of Earth's magnetic field. The notion of a conducting region was reinvoked by others, notably in 1902 by the American engineer Arthur E. Kennelly and the English physicist Oliver Heaviside, to explain the transmission of radio signals around the curve of Earth's surface before definitive evidence was obtained in 1925. For some years the ion-rich region was referred to as the Kennelly-Heaviside layer.

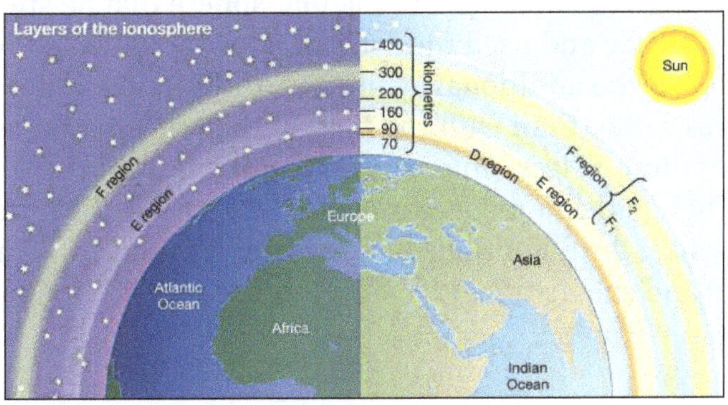

Layers of Earth's ionosphere: The day-and-night differences in the layers of Earth's ionosphere.

The name "ionosphere" was introduced first in the 1920s and was formally defined in 1950 by a committee of the Institute of Radio Engineers as "the part of the earth's upper atmosphere where ions and electrons are present in quantities sufficient to affect the propagation of radio waves." Much of the early research on the ionosphere was carried out by radio engineers and was stimulated by the need to define the factors influencing long-range radio communication. Subsequent research has focused on understanding the ionosphere as the environment for Earth-orbiting satellites and, in the military arena, for ballistic missile flight. Scientific knowledge of the ionosphere has grown tremendously, fueled by a steady stream of data from spacecraft-borne instruments and enhanced by measurements of relevant atomic and molecular processes in the laboratory.

The British Ariel 4 (U.K. 4), shown suspended with its solar panels deployed in flight configuration. It was launched Dec. 11, 1971, by NASA (National Aeronautics and Space Administration) and surveyed the ionosphere and certain radio signals.

Layers of the Ionosphere

Historically, the ionosphere was thought to be composed of a number of relatively distinct layers that were identified by the letters D, E, and F. The F layer was subsequently divided into regions F_1 and F_2. It is now known that all these layers are not particularly distinct, but the original naming scheme persists.

It appears that Edward V. Appleton, a pioneer in early radio probing of the ionosphere, is responsible for the nomenclature. Appleton was accustomed to using the symbol E to describe the electric field of the wave reflected from the first layer of the ionosphere that he studied. Later he identified a second layer at higher altitude and used the symbol F for the reflected wave. Suspecting a layer at lower altitude, he adopted the additional symbol D. In time, the letters came to be associated with the layers themselves rather than with the field of the reflected waves. It is now known that electron density increases more or less uniformly with altitude from the D region, reaching a maximum in the F_2 region. Though the nomenclature used to describe the different layers of the ionosphere continues in wide use, the definitions have evolved to reflect the improved understanding of the underlying physics and chemistry.

D Region

The D region is the lowest ionospheric region, at altitudes of about 70 to 90 km (40 to 55 miles). The D region differs from the E and F regions in that its free electrons almost totally disappear during the night, because they recombine with oxygen ions to form electrically neutral oxygen molecules. At this time, radio waves pass through to the strongly reflecting E and F layers above. During the day some reflection can be obtained from the D region, but the strength of radio waves is reduced; this is the cause of the marked reduction in the range of radio transmissions in daytime. At its upper boundary the D region merges with the E region.

E Region

The E region is also called Kennelly-Heaviside layer, named for American electrical engineer Arthur E. Kennelly and English physicist Oliver Heaviside in 1902. It extends from an altitude of 90 km (60 miles) to about 160 km (100 miles). Unlike that of the D region, the ionization of the E region remains at night, though it is considerably diminished. The E region was responsible for the reflections involved in Guglielmo Marconi's original transatlantic radio communication in 1902. The ionization density is typically 10^5 electrons per cubic centimetre during the day, though intermittent patches of stronger ionization are sometimes observed.

F Region

The F region extends upward from an altitude of about 160 km (100 miles). This region has the greatest concentration of free electrons. Although its degree of ionization persists with little change through the night, there is a change in the ion distribution. During the day, two layers can be distinguished: a small layer known as F_1 and above it a more highly ionized dominant layer called F_2. At night they merge at about the level of the F_2 layer, which is also called the Appleton layer. This region reflects radio waves with frequencies up to about 35 megahertz; the exact value depends on the peak amount of the electron concentration, typically 10^6 electrons per cubic centimetre, though with large variations caused by the sunspot cycle.

Mechanisms of Ionization

Photoionization

Most of the electrical activity in the ionosphere is produced by photoionization (ionization caused by light energy). Photons of short wavelength (that is, of high frequency) are absorbed by atmospheric gases. A portion of the energy is used to eject an electron, converting a neutral atom or molecule to a pair of charged species—an electron, which is negatively charged, and a companion positive ion. Ionization in the F1 region is produced mainly by ejection of electrons from molecular oxygen (O_2), atomic oxygen (O), and molecular nitrogen (N_2). The threshold for ionization of O_2 corresponds to a wavelength of 102.7 nm (nanometres, or billionths of a metre). Thresholds for O and N_2 are at 91.1 nm and 79.6 nm, respectively.

Positive ions in turn can react with neutral gases. There is a tendency for these reactions to favour production of more-stable ions. Thus, ionized atomic oxygen, O^+, can react with O_2 and N_2, resulting in ionized molecular oxygen (O_2^+) and ionized nitric oxide (NO^+), as shown by:

$$O^+ + O_2 \rightarrow O + O_2^+$$

and

$$O^+ + N_2 \rightarrow NO^+ + N.$$

Similarly, ionized molecular nitrogen (N_2^+) can react with O and O_2 to form NO^+ and O_2^+ as follows:

$$N_2^+ + O \rightarrow NO^+ + N$$

and

$$N_2^+ + O_2 \rightarrow N_2 + O_2^+$$

The most stable, and consequently most abundant, ions in the E and F_1 regions are O_2^+ and NO^+, the latter more so than the former. At lower altitudes, O_2^+ can react with the minor species of atomic nitrogen (N) and nitric oxide (NO) to form NO^+, as indicated by:

$$O_2^+ + N \rightarrow O + NO^+$$

and

$$O_2^+ + NO \rightarrow O_2 + N_2^+$$

In the D region, NO^+ and water vapour (H_2O) can interact to form the hydronium ion, H_3O^+, and companion species such as $H_5O_2^+$ and $H_7O_4^+$. Production of hydrated ions is limited by the availability of H_2O. As a consequence, they are confined to altitudes below about 85 km (53 miles).

Recombination

The electron density in the D, E, and F_1 regions reflects for the most part a local balance between production and loss. Electrons are removed mainly by dissociative recombination, a process in which electrons attach to positively charged molecular ions and form highly energetic, unstable neutral molecules. These molecules decompose spontaneously, converting internal energy to kinetic energy possessed by the fragments. The most important processes in the ionosphere involve recombination of O_2^+ and NO^+. These reactions may be summarized by:

$$O_2^+ + e \rightarrow O + O$$

and

$$NO^+ + e \rightarrow N + O$$

A portion of the energy released in reactions ($O_2^+ + e \rightarrow O + O$) and ($NO^+ + e \rightarrow N + O$) may appear as internal excitation of either nitrogen, oxygen, or both. The excited atoms can radiate, emitting faint visible light in the green and red regions of the spectrum, contributing to the phenomenon of airglow. Airglow originates mainly from altitudes above 80 km (50 miles) and is responsible for the diffuse background light that makes it possible to distinguish objects at Earth's surface on dark, moonless nights. Airglow is produced for the most part by reactions involved in the recombination of molecular oxygen. The contribution from reactions ($O_2^+ + e \rightarrow O + O$) and ($NO^+ + e \rightarrow N + O$) is readily detectable, however, and provides a useful technique with which to observe changes in the ionosphere from the ground. Over the years, studies of airglow have contributed significantly to scientific understanding of processes in the upper atmosphere.

As indicated above, dissociative recombination provides an effective path for removal of molecular ions. There is no comparable means for removal of atomic ions. Direct recombination of ionized atomic oxygen (O^+) with an electron requires that the excess energy be radiated as light. Radiative recombination is inefficient, however, compared with dissociative recombination and plays only a small role in the removal of ionospheric electrons. The situation becomes more complicated at high altitudes where atomic oxygen (O) is the major constituent of the neutral atmosphere and where electrons are produced primarily by its photoionization. The atomic oxygen ion, O^+, may react with N_2 and O_2 to form NO^+ and O_2^+, but the abundances of N_2 and O_2 decline relative to O as a function of increasing altitude. In the absence of competing reactions, the concentration of O^+ and the density of electrons would increase steadily with altitude, paralleling the rise in the relative abundance of O. This occurs to some extent but is limited eventually by vertical transport.

Diffusion

Ions and electrons produced at high altitude are free to diffuse downward, guided by Earth's magnetic field. The lifetime of O^+ is long at high altitudes, where the densities of O_2 and N_2 are very small. As ions move downward, the densities of O_2 and N_2 increase. Eventually the time constant for reaction of O^+ with O_2 and N_2 becomes comparable to the time for diffusion, and O^+ reacts to produce either O_2^+ or NO^+ before it can move much farther. The O^+ density exhibits a maximum in this region. Competition between chemistry and transport is responsible for the formation of an electron-density maximum in the F_2 layer. The dominant positive ion is O^+.

The density of O^+ decreases with decreasing altitude below the peak, reflecting a balance between production of O by photoionization and its removal by reactions ($O^+ + O_2 \rightarrow O + O_2^+$) and ($O^+ + N_2 \rightarrow NO^+ + N$). The density of O^+ also decreases above the peak. In this case, removal of photo-ions is regulated by downward diffusion rather than by chemistry. The distribution of O^+ with altitude above the peak reflects a balance of forces—a pressure-gradient force that acts to support O^+ in opposition to gravitational and electrostatic forces that combine to pull O^+ down. The electrostatic force acts to preserve electrical charge neutrality. In its absence, the concentration of ions—which are much more massive than electrons—would tend to fall off more rapidly with altitude than electrons. The abundance of electrons would quickly exceed that of ions, and the upper atmosphere would accumulate negative charge. The electric field redresses the imbalance by drawing electrons down and providing additional upward support for positively charged ions. Though O^+ has a mass of 16 atomic units, its abundance decreases with altitude as if it had a mass of only 8 atomic units. (One atomic unit corresponds to the mass of a hydrogen atom, 1.66 10^{-24} gram.) This discrepancy occurs because the electric field exerts a force that is equivalent to that exerted by the gravitational force on a body with a mass of eight atomic units. This electrostatic force is directed upward for ions and downward for electrons, in effect buoying the ions while encouraging the electrons to sink. The concentration of electrons therefore falls off with altitude at precisely the same rate as that of O^+, preserving the balance of positive and negative charge.

Photon Absorption

Ionization at any given level depends on three factors—the availability of photons of a wavelength capable of effecting ionization, a supply of atoms and molecules necessary to intercept this radiation, and the efficiency with which the atoms and molecules are able to do so. The efficiency is relatively large for O, O_2, and N_2 from about 10 to 80 nm. This is the portion of the spectrum responsible for production of electrons and ions in the F1 region. Photons with wavelengths between 90 and 100 nm are absorbed only by O_2. They therefore penetrate deeper and are responsible for producing about half the ionization in the E layer. The balance is derived from so-called "soft"

X-rays (those of longer wavelengths), which are absorbed with relatively low efficiency in the F region and so are able to penetrate to altitudes of about 120 km (75 miles) when the Sun is high over the region. "Hard" X-rays (those of shorter wavelengths—that is, below about 5 nm) reach even deeper. This portion of the spectrum accounts for the bulk of the ionization in the D region, with an additional contribution from wavelengths longer than 102.6 nm—mainly from photons in the strong solar emission line at Lyman α at a wavelength of 121.7 nm. (The Lyman series is a related sequence of wavelengths that describe electromagnetic energy given off by energized atoms in the ultraviolet region.) Lyman α emissions are weakly absorbed by the major components of the atmosphere—O, O_2, and N_2—but they are absorbed readily by NO and have sufficient energy to ionize this relatively unstable compound. Despite the low abundance of NO, the high flux of solar radiation at Lyman α is able to provide a significant source of ionization for the D region near 90 km (55 miles).

Ionospheric Variations

The ionosphere is variable in space and time. Some of the changes are chemical in origin and can be readily understood on the basis of the general considerations outlined above. There is a systematic variation, for example, according to the time of day. In early morning the Sun is relatively low in the sky, so that radiation must penetrate a large column of air before reaching a given level of the atmosphere. As a result, ionization rates are lower, and the location of ionized layers shifts to higher altitudes. As the Sun rises, the D, E, and F_1 layers shift in altitude. The layers are lowest and densities of electrons are highest at noon. At night, on the other hand, ionization in the D, E, and F_1 regions tends to disappear as electrons and ions recombine to form neutral gases.

The diurnal, or daily, variation of the F_2 layer is less dramatic. Ions produced at high altitudes during the day maintain a sizable density of electrons at the F_2 peak throughout the day and then diffuse downward at night. This accounts for the fact that radio reception (both in the broadcast and shortwave bands) is generally best at night. Ionization at lower altitudes—primarily those corresponding to the D region—tends to interfere with radio transmissions during the day. Interference is minimal at night because ionization in the D layer effectively disappears with the setting of the Sun.

The density of ionization varies in response to changes in the intensity and properties of radiation from the Sun. The output of solar energy is relatively constant in the visible and near-ultraviolet portions of the spectrum. It varies appreciably, however, at shorter wavelengths, reflecting changes in the temperature of the outermost regions of the solar atmosphere. The changes are particularly large, in excess of a factor of 10, at X-ray wavelengths. Variations in the D region are correspondingly large, with smaller though still significant changes in the E and F layers.

Solar activity varies on a characteristic timescale of about 11 years. It is not entirely periodic, however; successive cycles can differ significantly, and there are indications that activity can be low for centuries. The Sun was quiet for more than 200 years from about 1600 to about 1850. Solar activity was particularly intense in 1958.

Ionization above the F_2 peak is removed mainly by downward diffusion of ions and electrons. Ions are constrained, however, to move along the magnetic field. The field is oriented horizontally at the magnetic equator, which is equidistant between the magnetic North Pole and the magnetic South

Pole, so vertical diffusion is inhibited at low latitudes. The density of ionized atomic oxygen (O^+) and electrons at low latitudes is therefore controlled by chemistry to a larger extent than at high latitudes. The F_2 peak is correspondingly higher in altitude, and the density of electrons is elevated accordingly.

Ions and electrons formed at high altitudes and low latitudes are transported to higher latitudes by thermospheric winds. As a result, the highest density of electrons at the F_2 peak is observed at intermediate latitudes, offset from the magnetic equator by about 10 degrees.

Transport can also affect the distribution of ionization at lower altitudes. The diurnal pattern of heating in the troposphere and stratosphere excites a spectrum of waves, some of which are free to propagate vertically. The amplitude of the waves grows significantly as the disturbance enters regions of lower density. Passage of the waves is associated with strong alternating horizontal winds. Ionization can be driven up inclined magnetic field lines at one altitude, while winds blowing in an opposite direction at higher altitudes can induce simultaneous downward motion. This can lead to a bunching of ionization—a local enhancement of the electron density. The mechanism is particularly important in the E region and is responsible for the phenomenon known as sporadic E.

The buildup of ionization is normally limited by dissociative recombination of molecular ions. At D and E region altitudes, however, the ionosphere contains a small but variable concentration of atomic ions, derived from ionization of metals ablated from meteorites. The density of metallic ions—notably those of sodium (Na^+), magnesium (Mg^{++}), and potassium (K^+)—is sometimes high enough to supply a layer of ionization with a density comparable to that of the F layer. This can result in a major temporary disruption of radio communications.

Winds generated in the lower ionosphere by thermal forcing from below have characteristic periods expressed as submultiples of a day. Waves with a period of 24 hours dominate at low latitudes, whereas those with a characteristic period of 12 hours are more important at high latitudes. The origin of the waves is basically similar to that of oceanic tides caused by the pull of lunar gravity. The vertical motion that generates ionospheric waves, however, is the result of the diurnal pattern of heating and cooling rather than gravity. Additional waves can arise owing to irregular forcing, associated, for example, with thunderstorms, motion over mountain ranges, and other small-scale meteorological disturbances. These small-scale disturbances are referred to as gravity waves to distinguish them from the more regular planetary-scale motions excited by the diurnal cycle of heating and cooling. The regular response to thermal forcing is known as the atmospheric tide.

Tides and gravity waves have similar effects on ionization in the E region. They both are responsible for concentrating ionization in layers. In combination with the large-scale system of winds in the lower thermosphere, they are also effective in driving an irregular current that flows in the E and lower F regions of the ionosphere. The current owes its origin to differences in the facility with which motions of ions and electrons are constrained by the magnetic field. It is associated with an electric field and results in a modulation of the magnetic field that can be readily detected at the surface. The current is particularly intense in the equatorial region, where it is known as the electrojet. The region of strong current flow is known as the dynamo region.

Protons (H^+) and helium ions (He^+) are important components of the ionosphere above the F_2 peak. They increase in abundance relative to ionized atomic oxygen (O^+) with increasing altitude.

Protons are produced by photoionization of atomic hydrogen (H),

$$hv + H \rightarrow H^+ + e$$

and by charge transfer from O^+ to H,

$$O^+ + H \rightarrow O + H^+.$$

Helium ions are formed by photoionization of helium. The distribution of H^+ and He^+ with altitude reflects the influence of the polarization electric field set up to preserve charge neutrality. When O^+ is the dominant ion, the polarization field acts to lift H^+ and He^+ with a force equivalent, but in opposite direction, to that exerted by the gravitational field on a particle with a mass of eight atomic units, as described in the section on Diffusion. Protons behave as though they have an effective gravitational mass of −7 atomic units (−7 = 1 − 8). The effective mass of He^+ is −4 atomic units (−4 = 4 − 8).

The abundance of H^+ and He^+ increases with altitude. Eventually H^+ becomes the dominant component of the outermost ionosphere, which is sometimes referred to as the protonosphere. The more uniform composition of the atmosphere at this level causes a reduction in the polarization field to one equivalent to the gravitational force acting on a body with a mass of 0.5 atomic unit, directed upward for ions and downward for electrons. This field is sufficient to maintain equal densities of H^+ and electrons. The effective masses of O^+ and He^+ shift to 15.5 atomic units (15.5 = 16 − 0.5) and 3.5 atomic units (3.5 = 4 − 0.5), respectively, and the abundance of O^+, He^+, and H^+ declines with further increases in altitude.

Magnetosphere

The overall structure of the outer ionosphere—the magnetosphere—is strongly influenced by the configuration of Earth's magnetic field. Close to the planet's surface, the magnetic field has a structure similar to that of an ideal dipole. Field lines are oriented more or less vertically at high latitudes, sweep back over the Equator, where they are essentially horizontal, and connect to Earth in a symmetrical pattern at high latitudes. The field departs from this ideal dipolar configuration, however, at high altitudes. There the terrestrial field, Earth's magnetic field, is distorted to a significant extent by the solar wind, with its embedded solar magnetic field. Ultimately the terrestrial field is dominated by the interplanetary field, which is generated by the Sun.

The Van Allen radiation belts contained within Earth's magnetosphere. Pressure from the solar wind is responsible for the asymmetrical shape of the magnetosphere and the belts.

The Van Allen radiation belts contained within Earth's magnetosphere. Pressure from the solar wind is responsible for the asymmetrical shape of the magnetosphere and the belts.

The solar wind compresses the magnetic field on Earth's dayside at a distance of about 10 Earth radii, or almost 65,000 km (40,000 miles) from the planet. At this distance the magnetic field is so weak that the pressure associated with particles escaping from Earth's gravity is comparable to the opposing pressure associated with the solar wind. This equilibrium region, with a characteristic thickness of 100 km (60 miles), is called the magnetopause and marks the outer boundary of the magnetosphere. The lower boundary of the magnetosphere is several hundred kilometres above Earth's surface.

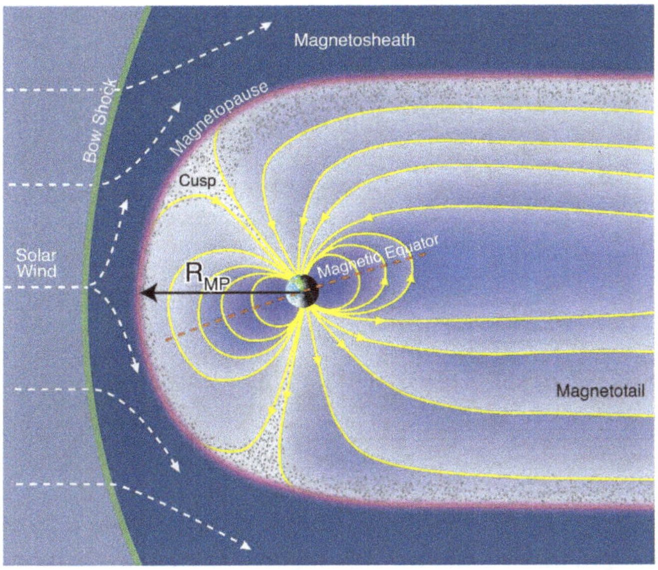

On the nightside, the terrestrial field is stretched out in a giant tail that reaches past the orbit of the Moon, extending perhaps to distances in excess of 1,000 Earth radii. The magnetotail can extend to such great distances because on the nightside the forces associated with the magnetic field and the solar wind are parallel.

The outermost regions of the magnetosphere are exceedingly complex, especially at high latitudes, where terrestrial field lines are open to space. Ionization from the solar wind can leak into the magnetosphere in a number of ways. It can enter by turbulent exchange at the dayside magnetopause or more directly at cusps in the magnetopause at high latitudes where closed loops of the magnetic field on the dayside meet fields connecting to the magnetotail. In addition, it can enter at large distances on the nightside, where the magnetic pressure is relatively low and where field lines can reconnect readily, providing easy access to the giant plasma sheet in the interior of Earth's magnetotail.

The magnetosheath, a region of magnetic turbulence in which both the magnitude and the direction of Earth's magnetic field vary erratically, occurs between 10 and 13 Earth radii toward the Sun. This disturbed region is thought to be caused by the production of magnetohydrodynamic shock waves, which in turn are caused by high-velocity solar wind particles. Ahead of this bow shock boundary, toward the Sun, is the undisturbed solar wind.

Auroras

Auroras are perhaps the most spectacular manifestations of the complex interaction of the solar wind with the outer atmosphere. The energetic electrons and protons responsible for an aurora are directed by the solar wind along magnetic fields into Earth's magnetosphere.

Auroras occur in both hemispheres, confined for the most part to high latitudes in oval-shaped regions that maintain a more or less fixed orientation with respect to the Sun. The centre of the auroral oval is displaced a few degrees to the nightside with respect to the geomagnetic pole. The midnight portion of the oval is, on average, at a geomagnetic latitude of 67°; the midday portion is at about 76°. An observer between 67° and 74° magnetic latitudes generally encounters auroras twice a day—once in evening and once in morning.

A display of aurora australis, or southern lights, manifesting itself as a glowing loop, in an image of part of Earth's Southern Hemisphere taken from space by astronauts aboard the U.S. space shuttle orbiter Discovery on May 6, 1991. The mostly greenish blue emission is from ionized oxygen atoms at an altitude of 100–250 km (60–150 miles). The red-tinged spikes at the top of the loop are produced by ionized oxygen atoms at higher altitudes, up to 500 km (300 miles).

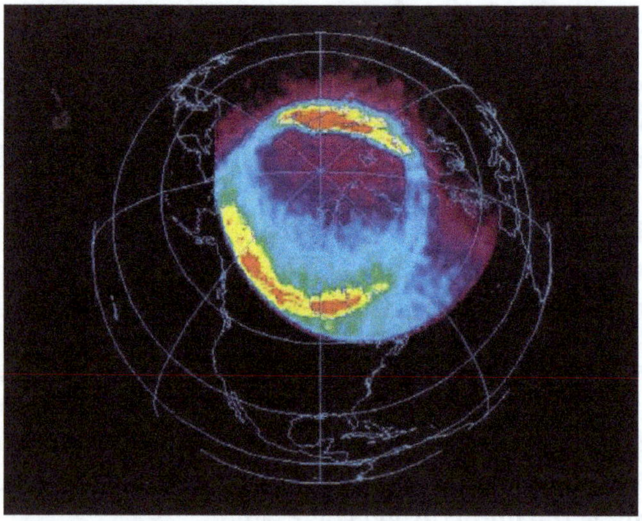

Earth's full North Polar auroral oval, in an image taken in ultraviolet light by the U.S. Polar spacecraft over northern Canada, April 6, 1996. In the colour-coded image, which simultaneously shows dayside and nightside auroral activity, the most intense levels of activity are red, and the lowest levels are blue. Polar, launched in February 1996, was designed to further scientists' understanding of how plasma energy contained in the solar wind interacts with Earth's magnetosphere.

Auroral Zones

The portion of Earth that traverses the midnight portion of the auroral oval is known as the auroral zone. In the Northern Hemisphere this zone lies along a curve extending from the northern regions of Scandinavia through Iceland, the southern tip of Greenland, the southern region of

Hudson Bay, central Alaska, and on to the coast of Siberia. This is the prime region from which to view an aurora in the Northern Hemisphere. The phenomenon is by no means static, however. The auroral zone shifts poleward at times of low solar activity, while during periods of high solar activity it has been known to move as far south as 40° (geographic latitude). At low latitudes, an aurora assumes a characteristic red colour. In ancient times this colour was often interpreted as evidence of impending disaster. More recently it has been taken as a sign of approaching fires. Auroras assume a variety of forms, depending on the vantage point from which they are observed. The luminosity of an aurora is generally aligned with the magnetic field. Field lines are close to vertical in polar regions, and so an aurora occurring there appears to stand on end, hanging from the sky in great luminous drapes. It is a spectacular sight indeed, especially if viewed from a distance either from the north or south. At lower latitudes, the magnetic field lines are inclined with respect to the vertical. There an aurora appears as streamers radiating from the zenith. Such is the majesty of the aurora that no two displays are totally alike. Light can move rapidly across the sky on some occasions, and at other times it can appear to stand in place, flickering on and off.

Causes of Auroral Displays

The most common type of aurora is associated with bombardment of the atmosphere by electrons with energies of up to 10,000 electron volts. The energy source for these electrons originates ultimately from the Sun. It is propagated through space by the solar wind along bundled, ropelike magnetic fields that form temporarily between the Sun and Earth's magnetosphere, most probably to the plasma sheet. Energetic electrons enter the atmosphere along magnetic field lines. They produce a shower of secondary and tertiary electrons, approximately one for every 35 electron volts of energy in the primary stream. Primaries can propagate to altitudes as low as 100 km (60 miles). Most of the luminosity is produced, however, by low-energy secondary and tertiary electrons. Prominent emissions in the spectrum of this luminosity are associated with the red line of atomic oxygen at 633 nm, the green line of atomic oxygen at 558 nm, the first negative bands of ionized molecular nitrogen at 391 nm and 428 nm, and a host of emissions from atomic oxygen, molecular oxygen, ionized molecular oxygen, and molecular nitrogen. Many of these features are present also in the day and night airglow. They are most notable in auroras because of their intensity and the rapidity with which they switch on and off in response to changes in the flux and energy of incoming primaries. An aurora has a characteristic red colour if the energy of primaries is relatively low. Emission in this case is dominated by atomic oxygen and is confined for the most part to altitudes above 250 km (150 miles). If the energy of the primaries is high, an aurora has a greenish blue colour and extends downward to altitudes as low as 90 km (55 miles).

Auroral displays are also produced by bombardment of the atmosphere by energetic protons. Protons with energies of up to 200,000 electron volts are responsible for auroral activity in a diffuse belt that is equatorward of the main auroral zone. These protons can be detected from the ground by observation of Doppler-shifted radiation emitted by fast hydrogen atoms formed by charge transfer from atmospheric atoms and molecules. Protons also play a role at higher latitudes, especially at times following major solar flares. It is thought that the protons responsible for auroras at the polar caps are solar in origin. Associated energies may reach as high as one million electron volts, and particles may penetrate as deep as 80 km (50 miles). Polar cap auroras can provide a significant transient source of mesospheric and stratospheric nitric oxide (NO). They can be responsible for small but detectable short-term fluctuations in the abundance of stratospheric ozone.

Van Allen Radiation Belts

The magnetosphere includes two doughnut-shaped radiation belts, or zones, centred on the Equator that are occupied by appreciable numbers of energetic protons and electrons trapped in the outermost reaches of the atmosphere. No real gap exists between the two zones; they actually merge gradually, with the flux of charged particles showing two regions of maximum density. The inner belt extends from roughly 1,000 to 5,000 km (600 to 3,000 miles) above the terrestrial surface and the outer belt from some 15,000 to 25,000 km (9,300 to 15,500 miles). The belts were named in honour of James A. Van Allen, the American physicist who discovered them in 1958. His was a triumph of serendipity—he detected the presence of the trapped particles with a Geiger counter designed to measure the flux of cosmic rays in space. It was the first great discovery of the space age and was achieved by combining data obtained with instruments carried by three of the earliest United States scientific satellites—Explorer 1, Explorer 4, and Pioneer 3.

The flux of protons crossing a square centimetre of surface in the inner Van Allen belt can be as large as 20,000 per second, higher than the flux of cosmic radiation in space by a factor of 10,000. Protons in the inner belt have energies in excess of 7 × 108 electron volts, enough to enable them to penetrate about 10 cm (4 inches) of lead. Spacecraft flying through the belts must be protected; otherwise, their electronic components would be subjected to irreparable damages.

The high-energy protons in the inner Van Allen belt are thought to originate from the decay of neutrons that are produced by the interaction of the atmosphere with energetic cosmic rays of galactic origin. Some of these short-lived neutrons—they have a lifetime of 12 minutes—are ejected upward. A fraction of them decay into energetic protons and electrons as they pass through the region occupied by the Van Allen belts. These protons and electrons become trapped and travel in spiral paths along the flux lines of Earth's magnetic field. The particles reverse their direction at intermediate altitudes (about 500 km [300 miles]) and low latitudes because, as the particles approach either of the magnetic poles, the increase in the strength of the field causes them to be reflected back toward the other pole. Collisions with atoms in the thin atmosphere eventually remove the particles from the belts, but they generally survive for about 10 years. This relatively long lifetime allows particles to accumulate in the radiation belts, providing high fluxes despite the small magnitude of the intrinsic source.

The inner belt merges gradually with the outer belt, which extends from about two to eight Earth radii. A portion of the ionization in the outer belt is derived from the solar wind, as demonstrated by the presence of helium ions in addition to protons. Unlike the outer zone, the inner belt contains no helium ions, while it has been established that helium ions account for about 10 percent of solar wind. The flux of electrons in the outer belt can vary by orders of magnitude over intervals as short as a few days. These changes appear to correlate with times of strong magnetic disturbances. They are not, however, as yet well understood.

OZONOSPHERE

Ozone layer, also called ozonosphere is the region of the upper atmosphere, between roughly 15 and 35 km (9 and 22 miles) above Earth's surface, containing relatively high concentrations of ozone molecules (O_3). Approximately 90 percent of the atmosphere's ozone occurs in the stratosphere, the region extending from 10–18 km (6–11 miles) to approximately 50 km (about 30 miles) above Earth's surface. In the stratosphere the temperature of the atmosphere rises with increasing height, a phenomenon created by the absorption of solar radiation by the ozone layer. The ozone layer effectively blocks almost all solar radiation of wavelengths less than 290 nanometres from reaching Earth's surface, including certain types of ultraviolet (UV) and other forms of radiation that could injure or kill most living things.

Location in Earth's Atmosphere

In the midlatitudes the peak concentrations of ozone occur at altitudes from 20 to 25 km (about 12 to 16 miles). Peak concentrations are found at altitudes from 26 to 28 km (about 16 to 17 miles) in the tropics and from about 12 to 20 km (about 7 to 12 miles) toward the poles. The lower height of the peak-concentration region in the high latitudes largely results from poleward and downward atmospheric transport processes that occur in the middle and high latitudes and the reduced height of the tropopause (the transition region between the troposphere and stratosphere).

Most of the remaining ozone occurs in the troposphere, the layer of the atmosphere that extends from Earth's surface up to the stratosphere. Near-surface ozone often results from interactions between certain pollutants (such as nitrogen oxides and volatile organic compounds), strong sunlight, and hot weather. It is one of the primary ingredients in photochemical smog, a phenomenon that plagues many urban and suburban areas around the world, especially during the summer months.

Ozone damage on the leaf of an English walnut.

Ozone Creation and Destruction

The production of ozone in the stratosphere results primarily from the breaking of the chemical bonds within oxygen molecules (O_2) by high-energy solar photons. This process, called photodissociation, results in the release of single oxygen atoms, which later join with intact oxygen molecules to form ozone. Rising atmospheric oxygen concentrations some two billion years ago allowed ozone to build up in Earth's atmosphere, a process that gradually led to the formation of the stratosphere. Scientists believe that the formation of the ozone layer played an important role in the development of life on Earth by screening out lethal levels of UVB radiation (ultraviolet radiation with wavelengths between 315 and 280 nanometres) and thus facilitating the migration of life-forms from the oceans to land.

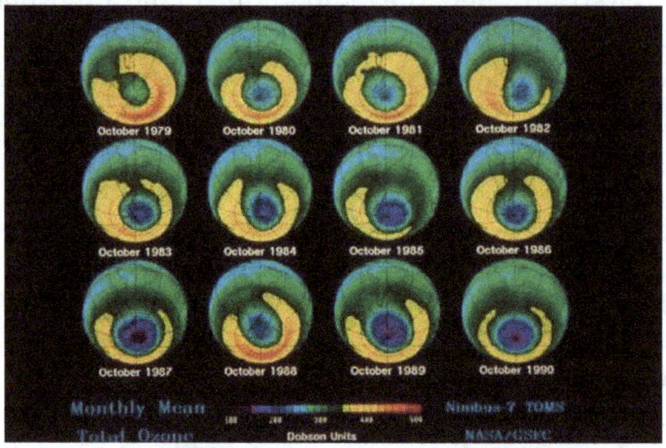

Ozone: Hole Changes in the size of the ozone hole from October 1979 to October 1990.

The amount of ozone in the stratosphere varies naturally throughout the year as a result of chemical processes that create and destroy ozone molecules and as a result of winds and other transport processes that move ozone molecules around the planet. Over the course of several decades, however, human activities substantially altered the ozone layer. Ozone depletion, the global decrease in stratospheric ozone observed since the 1970s, is most pronounced in polar regions, and it is well correlated with the increase of chlorine and bromine in the stratosphere. Those chemicals, once freed by UV radiation from the chlorofluorocarbons (CFCs) and other halocarbons (carbon-halogen compounds) that contain them, destroy ozone by stripping away single oxygen atoms from ozone molecules. Depletion is so extensive that so-called ozone holes (regions of severely reduced ozone coverage) form over the poles during the onset of their respective spring seasons. The largest such hole—which has spanned more than 20.7 million square km (8 million square miles) on a consistent basis since 1992—appears annually over Antarctica between September and November.

Ozone depletion: Antarctic ozone hole.

As the amount of stratospheric ozone declines, more UV radiation reaches Earth's surface, and scientists worry that such increases could have significant effects on ecosystems and human health. The concern over exposure to biologically harmful levels of UV radiation has been the main driver of the creation of international treaties such as the Montreal Protocol on Substances That Deplete the Ozone Layer and its amendments, designed to protect Earth's ozone layer. Compliance with international treaties that phased out the production and delivery of many ozone-depleting chemicals, combined with upper stratospheric cooling due to increased carbon dioxide, is thought to have contributed to the shrinking of the ozone holes over the poles and to slightly higher stratospheric ozone levels overall. Continued reductions in chlorine loading are expected to result in smaller ozone holes above Antarctica after 2040. However, some scientists noted that gains in stratospheric ozone levels have only occurred in the upper stratosphere, with declines in ozone concentrations in the lower stratosphere outpacing increases in the upper stratosphere.

References

- Atmosphere-layers: ucar.edu, Retrieved 21 March, 2019
- Troposphere, atmosphere: enviropedia.org.uk, Retrieved 12 April, 2019
- Stratosphere-overview, shortcontent: ucar.edu, Retrieved 23 June, 2019
- Mesosphere-facts, weather, facts: softschools.com, Retrieved 15 May, 2019
- Thermosphere-overview, shortcontent: ucar.edu, Retrieved 20 July, 2019
- Exosphere-definition-and-facts: thoughtco.com, Retrieved 25 August, 2019
- Ionosphere-and-magnetosphere, science: britannica.com, Retrieved 31 July, 2019
- Ozone-layer, science: britannica.com, Retrieved 9 May, 2019

4

Atmospheric Phenomena

Atmospheric phenomena are the observable events that are the outcome of the interactions of light and matter. It includes phenomena like belt of venus, green flash, 22 degree halos, thunderstorm, jet stream, etc. All the diverse principles of these atmospheric phenomena have been carefully analyzed in this chapter.

GREEN FLASH

A green flash, also referred to as an emerald flash, appears at the tip top of the sunrise just before it breaches the horizon or it is seen as the last hint of the sun setting on the horizon. It is most commonly reported at sunset, but is actually just as possible to view at either sunrise or sunset. Logically, it is easier to know when and where the sun is setting than to catch the time and place of the sun rising. Even more rare is a green ray which is a green flash accompanied by a shaft of light shooting upwards from the top of the sun.

During that split second is when the momentary flash of green appears. Some have reported seeing more than one green flash less than a second apart from each other. It is also possible to see it when the sun is rising or setting over a mountain or even a cloud. However, the most common place to see a green flash is at sea when the wide open water and clear skies offer a horizon ideal for catching this rare atmospheric phenomenon. Basically, it is best seen when you have a distant and distinct cloud-free horizon.

Why does the top of the sun appear green? Think of the sunset and how the sun's orange and red colors look hazy and huge as the sun appears to sink and touch the horizon. The reds of a sunset are caused by the same effect as a green flash—both are caused by refraction of light. As the sun sets, the light is viewed through a greater and greater density of molecules and the light is therefore refracted as the atmospheric soup acts as a prism spreading the rainbow of light. As the light passes through water vapor and other particles in the atmosphere this prism effect causes the sunlight to absorb and refract different wavelengths of electromagnetic energy we call colors. This explains how the oceans horizon with its thermal difference of water and air create an ideal setting for a mirage. At the last seconds of a sunset the color green in the light spectrum is refracted enough to give off the mirage of green. The green appears separate, just above the red-orange setting sunlight. This optical mirage, as seen by the human eye, is a case of looking at the right place at the right time under the right conditions.

Other Objects Where These Phenomena can be Seen

Interestingly, if a planet such as Venus or Jupiter is observed through a telescope, viewed as it sets below the horizon, a greenish refracted light effect is seen. The setting planet appears to turn reddish orange as it sets on the horizon, just as the sun does, and at the last seconds appears greenish at its tip top. This visual effect lasts for a 20 arc second band and for about 1.4 seconds of time as measured by atmospheric optics. Due to the differences in the way a lens takes in light and how the human eye takes in light, this effect can appear differently, more clear and distinct, when captured by a camera with a telescopic lens.

It is said that once you seen a green flash you will never go wrong in matters of the heart. This originates largely from Jules Verne's 1882 novel "Le Rayon Vert" (The Green Ray). When something in nature is so rare and beautiful as a green flash, or better yet a brilliant green ray, humankind often attributes meaning and legends. Since it is most common on the high seas, tales of horizon gazing sailors passed on tales of the green flashes given off by the sun. Most humans today do not take the time to gaze at a sunset or the patience and focus to await the rising sun. There are potentially two chances each day! Good luck in the search for the mysterious green flash.

BELT OF VENUS

The "Belt of Venus" is an atmospheric phenomenon that creates a pink band in the sky at sunrise and sunset. It is actually the area between Earth's shadow and the blue sky. The belt is similar to alpenglow, which creates a reddish glow just over the horizon.

- The Belt of Venus is an atmospheric optical phenomenon, but could also be considered a space phenomenon.
- The pink color in the belt is sunlight that is shining through denser atmosphere near the Earth's surface, and being reflected off of atmosphere at the opposite end of the sky.
- The dark blue layer in the sky below the pink belt is Earth's "shadow" being cast onto the atmosphere.
- It is visible at sunrise or sunset, but is more pronounced at sunset.
- It is best seen during the summer months, on clear and slightly hazy evenings.
- The Belt of Venus has several names, including antitwilight arch.

What causes the Belt of Venus?

Several factors contribute to the Belt of Venus phenomenon.

- Anti-twilight or Alpenglow forms: During sunset on clear evenings, the atmosphere around the horizon opposite of the Sun appears as a light shade of orange and pink. The pink color is sunlight that is shining through denser atmosphere near Earth's western horizon, and reflecting off of atmosphere on the opposite side of the observer. The term for this reflection back towards the observer is backscattering. This creates a pink "band" near the eastern horizon. The same pink light is often seen on mountain tops at sunset, which is where the term "Alpenglow" comes from.

- Earth's shadow appears: As Earth rotates, sunlight ceases to reach parts of the atmosphere near Earth's horizon on the side of the sky opposite the Sun. Earth's shadow is being cast on the atmosphere, and it is no longer reflecting direct sunlight.
- Earth's shadow rises: As Earth's shadow continues to emerge, it makes the pink band appear to separate from the horizon. This gives the band more of a dramatic arch shape. We call this arch the Belt of Venus.

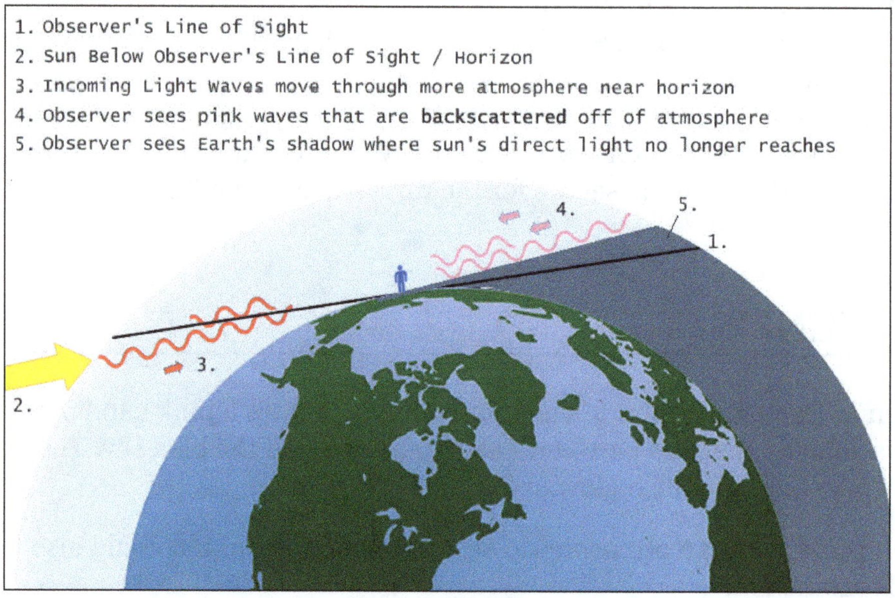

Belt of Venus.

Intensity of the Belt

Backscatter causes the colors of the belt to become more pronounced and vivid right before the sun rises or just after it sets for the day. During sunrise and sunset, the effects of scattering are magnified, which is why a rising and setting sun typically has a spectacular red glow.

Best Time to View the Belt

While this phenomenon can be seen at sunrise, the best time to find it is at sunset. As the Sun is setting in the West, the belt will be visible in the eastern sky (remember that it's visible opposite the sun). The best conditions for getting a nice view are when the sky is clear, when there's minimal wind and humidity, and when there are no obstructions on the horizon. When it's visible, the belt will appear anywhere from 10° to 46° over the horizon line.

Why is it Called the Belt of Venus?

It is called this due to an association of the phenomenon with the greek goddess Aphrodite and roman goddess Venus. It is not associated with the planet Venus.

22 DEGREE HALOS

The halo is the eye picking up light interacting with ice crystals in high-level cirrus and cirrostratus clouds as they pass in front of the sun or moon. These clouds are characterized as thin and wispy strands of ice that form in the cold region of the atmosphere, typically at around 18,000-21,000 feet or more above the surface of the planet.

This high, icy atmospheric region explains how the halo effect can be seen even in the heat of summer or over steamy tropical regions. Halos are brightest when the sun or moon are lower on the horizon rather than higher in the sky. Halos can also be refracted through near surface ice crystals not just through high flying clouds. In very cold climates where ice crystals form in the air near ground level, known as diamond dust, halos, sundogs and other optical effects of light appear.

The distance of the light of a halo averages about 22 degrees from the center of the sun or moon. This standard distance has to do with the way light from incoming sunlight changes angle by about 22 degrees as the sunlight exits the ice crystals. Sunlight hitting millions of plate-shaped hexagonal ice crystals causes this effect. There is no light refracted from angles less than 22 degrees due to the shape of the ice crystals. This is what gives the halo its dark center and what creates the sharp hit of light seen at the 22 degree point distance. The crystals have to be oriented in specific ways and have to be less than 20.5 micrometers across to create the effect.

Sundogs

Sundogs, also known as parhelia, are another brilliant optical effect associated with 22 degree halos. Sundogs are bright spots, sometimes called mock suns, seen as subtly colored patches of light to the right and left of the sun on or slightly outside a 22 degree halo circle. They are brightest and tallest when the sun in near the horizon at sunset or sunrise. Sundogs are caused by light interacting with hexagonal ice crystals just as with the light seen from a halo. Sundogs usually come in pairs, however sometimes only one sundog appears either to the left or right of the sun due to atmospheric conditions. Once the sun reaches about 40 degrees above the horizon sundogs are very difficult to see and fade from view.

The color of light in a halo and in sundogs shows red on the sharp, bright inside edge of the halo. The color changes to orange and yellow then bluish on the hazier outside edge of light. Strong sundogs give an amazing rainbow show, many halos and sundogs appear only as white light.

Moon Halos

The moon forms 22 degree halos and its dimmer, less colorful equivalent to sundogs are called paraselenae. These moon dogs, like sun dogs, form to the left and right of the moon along or near the halo. The best moon phase to see a moon halo is when the moon is full, however during other moon phases it is also possible.

When do Halos Appear?

In many cases the high sheets and isolated streaks of icy cirrus clouds means there is stable weather to come. However, a large number of approaching cirrus clouds can indicate a frontal system is moving in. In the tropics, a veil of cirrus clouds typically approaches as the visible leading edge of a hurricane. Sundogs and halos are caused by light passing through cirrus clouds and therefore do provide clues about the weather, they are however not reliable signs of wet or stormy weather to come.

Halos and sundogs are seen only by an eye -or a camera- that is looking from a specific angle at the right time. Someone standing only a few meters away, or on a hilltop away, would not see the halo the same or not see the halo at all. Someone standing at a different point would see the halo through a different set of ice crystals. A halo only exists if there is an eye to see it. Just as a rainbow can be viewed only from a particular angle and then disappears. This makes finding the pot of gold an elusive task.

THUNDERSTORM

Thunderstorm is a violent, short-lived weather disturbance that is almost always associated with lightning, thunder, dense clouds, heavy rain or hail, and strong, gusty winds. Thunderstorms arise when layers of warm, moist air rise in a large, swift updraft to cooler regions of the atmosphere. There the moisture contained in the updraft condenses to form towering cumulonimbus clouds and, eventually, precipitation. Columns of cooled air then sink earthward, striking the ground with strong downdrafts and horizontal winds. At the same time, electrical charges accumulate on cloud particles (water droplets and ice). Lightning discharges occur when the accumulated electric charge becomes sufficiently large. Lightning heats the air it passes through so intensely and quickly that shock waves are produced; these shock waves are heard as claps and rolls of thunder. On occasion, severe thunderstorms are accompanied by swirling vortices of air that become concentrated and powerful enough to form tornadoes.

Thunderstorms are known to occur in almost every region of the world, though they are rare in polar regions and infrequent at latitudes higher than 50° N and 50° S. The temperate and tropical regions of the world, therefore, are the most prone to thunderstorms. In the United States the areas of maximum thunderstorm activity are the Florida peninsula (more than 90 thunderstorm days per year), the Gulf Coast (70–80 days per year), and the mountains of New Mexico (50–60 days per year). Central Europe and Asia average 20 to 60 thunderstorm days per year. It has been estimated that at any one moment there are approximately 1,800 thunderstorms in progress throughout the world.

Thunderstorm Formation and Structure

Vertical Atmospheric Motion

Most brief but violent disturbances in Earth's wind systems involve large areas of ascending and descending air. Thunderstorms are no exception to this pattern. In technical terms, a thunderstorm is said to develop when the atmosphere becomes "unstable to vertical motion." Such an instability can arise whenever relatively warm, light air is overlain by cooler, heavier air. Under such conditions the cooler air tends to sink, displacing the warmer air upward. If a sufficiently large volume of air rises, an updraft (a strong current of rising air) will be produced. If the updraft is moist, the water will condense and form clouds; condensation in turn will release latent heat energy, further fueling upward air motion and increasing the instability.

Once upward air motions are initiated in an unstable atmosphere, rising parcels of warm air accelerate as they rise through their cooler surroundings because they have a lower density and are more buoyant. This motion can set up a pattern of convection wherein heat and moisture are transported upward and cooler and drier air is transported downward. Areas of the atmosphere where vertical motion is relatively strong are called cells, and when they carry air to the upper troposphere (the lowest layer of the atmosphere), they are called deep cells. Thunderstorms develop when deep cells of moist convection become organized and merge, and then produce precipitation and ultimately lightning and thunder.

Upward motions can be initiated in a variety of ways in the atmosphere. A common mechanism is by the heating of a land surface and the adjacent layers of air by sunlight. If surface heating is sufficient, the temperatures of the lowest layers of air will rise faster than those of layers aloft, and the air will become unstable. The ability of the ground to heat up quickly is why most thunderstorms form over land rather than oceans . Instability can also occur when layers of cool air are warmed from below after they move over a warm ocean surface or over layers of warm air. Mountains, too, can trigger upward atmospheric motion by acting as topographic barriers that force winds to rise. Mountains also act as high-level sources of heat and instability when their surfaces are heated by the Sun.

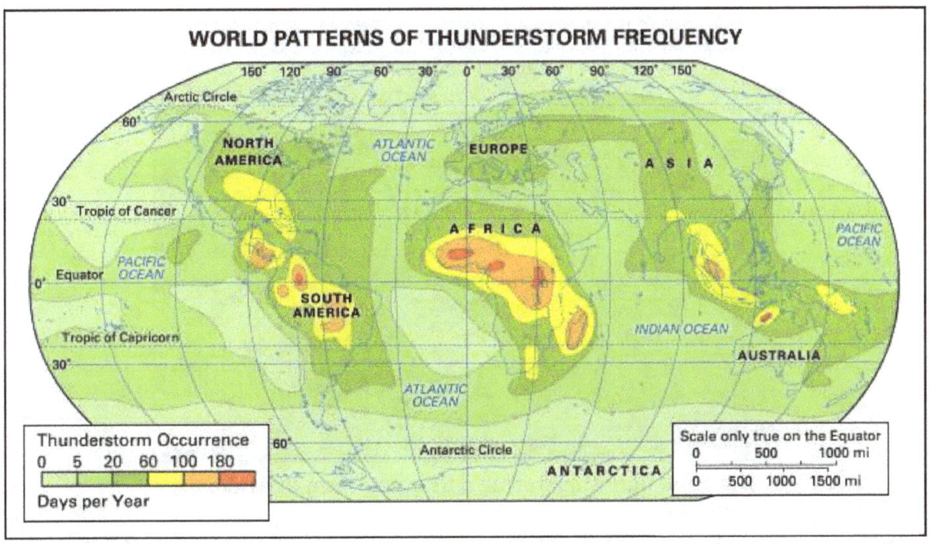

World patterns of thunderstorm frequency: Thunderstorms occur most often in the tropical latitudes over land, where the air is most likely to heat quickly and form strong updrafts.

The huge clouds associated with thunderstorms typically start as isolated cumulus clouds (clouds formed by convection, as described above) that develop vertically into domes and towers. If there is enough instability and moisture and the background winds are favourable, the heat released by condensation will further enhance the buoyancy of the rising air mass. The cumulus clouds will grow and merge with other cells to form a cumulus congestus cloud extending even higher into the atmosphere (6,000 metres [20,000 feet] or more above the surface). Ultimately, a cumulonimbus cloud will form, with its characteristic anvil-shaped top, billowing sides, and dark base. Cumulonimbus clouds typically produce large amounts of precipitation.

Structure of a Thunderstorm

When the atmosphere becomes unstable enough to form large, powerful updrafts and downdrafts (as indicated by the red and blue arrows), a towering thundercloud is built up. At times the updrafts are strong enough to extend the top of the cloud into the tropopause, the boundary between the troposphere (or lowest layer of the atmosphere) and the stratosphere.Click on the icons along the left-hand side of the figure to view illustrations of other phenomena associated with its characteristic anvil-shaped top, billowing sides, and dark base. Cumulonimbus clouds typically produce large amounts of precipitation.

Types of Thunderstorms

At one time, thunderstorms were classified according to where they occurred—for example, as local, frontal, or orographic (mountain-initiated) thunderstorms. Today it is more common to classify storms according to the characteristics of the storms themselves, and such characteristics depend largely on the meteorological environment in which the storms develop. The United States National Weather Service has defined a severe thunderstorm as any storm that produces a tornado, winds greater than 26 metres per second (94 km [58 miles] per hour), or hail with a diameter greater than 1.9 cm (0.75 inch).

Isolated Thunderstorms

Rain and lightning during a thunderstorm in Arizona.

Isolated thunderstorms tend to occur where there are light winds that do not change dramatically with height and where there is abundant moisture at low and middle levels of the atmosphere—that is, from near the surface of the ground up to around 10,000 metres (33,000 feet) in altitude. These storms are sometimes called air-mass or local thunderstorms. They are mostly vertical in structure, are relatively short-lived, and usually do not produce violent weather at the ground. Aircraft and radar measurements show that such storms are composed of one or more convective cells, each of which goes through a well-defined life cycle. Early in the development of a cell, the air motions are mostly upward, not as a steady, uniform stream but as one that is composed of a series of rising eddies. Cloud and precipitation particles form and grow as the cell grows. When the accumulated load of water and ice becomes excessive, a downdraft starts. The downward motion is enhanced when the cloud particles evaporate and cool the air—almost the reverse of the processes in an updraft. At maturity, the cell contains both updrafts and downdrafts in close proximity. In its later stages, the downdraft spreads throughout the cell and diminishes in intensity as precipitation falls from the cloud. Isolated thunderstorms contain one or more convective cells in different stages of evolution. Frequently, the downdrafts and associated outflows from a storm trigger new convective cells nearby, resulting in the formation of a multiple-cell thunderstorm.

Solar heating is an important factor in triggering local, isolated thunderstorms. Most such storms occur in the late afternoon and early evening, when surface temperatures are highest.

Multiple-cell Thunderstorms and Mesoscale Convective Systems

Violent weather at the ground is usually produced by organized multiple-cell storms, squall lines, or a supercell. All of these tend to be associated with a mesoscale disturbance (a weather system of intermediate size, that is, 10 to 1,000 km [6 to 600 miles] in horizontal extent). Multiple-cell storms have several updrafts and downdrafts in close proximity to one another. They occur in clusters of cells in various stages of development moving together as a group. Within the cluster one cell dominates for a time before weakening, and then another cell repeats the cycle. In squall lines, thunderstorms form in an organized line and create a single, continuous gust front (the leading edge of a storm's outflow from its downdraft). Supercell storms have one intense updraft and downdraft.

Sometimes the development of a mesoscale weather disturbance causes thunderstorms to develop over a region hundreds of kilometres in diameter. Examples of such disturbances include frontal wave cyclones (low-pressure systems that develop from a wave on a front separating warm and cool air masses) and low-pressure troughs at upper levels of the atmosphere. The resulting pattern of storms is called a mesoscale convective system (MCS). Severe multiple-cell thunderstorms and supercell storms are frequently associated with MCSs. Precipitation produced by these systems typically includes rainfall from convective clouds and from stratiform clouds (cloud layers with a large horizontal extent). Stratiform precipitation is primarily due to the remnants of older cells with a relatively low vertical velocity—that is, with limited convection occurring.

Lightning: Cloud-to-ground lightning discharge in a field from a cumulonimbus cloud.

Thunderstorms can be triggered by a cold front that moves into moist, unstable air. Sometimes squall lines develop in the warm air mass tens to hundreds of kilometres ahead of a cold front. The tendency of prefrontal storms to be more or less aligned parallel to the front indicates that they are initiated by atmospheric disturbances caused by the front.

In the central United States, severe thunderstorms commonly occur in the springtime, when cool westerly winds at middle levels (3,000 to 10,000 metres [10,000 to 33,000 feet] in altitude) move over warm and moist surface air flowing northward from the Gulf of Mexico. The resulting broad region of instability produces MCSs that persist for many hours or even days.

In the tropics, the northeast trade winds meet the southeast trades near the Equator, and the resulting intertropical convergence zone (ITCZ) is characterized by air that is both moist and unstable. Thunderstorms and MCSs appear in great abundance in the ITCZ; they play an important role in the transport of heat to upper levels of the atmosphere and to higher latitudes.

Supercell Storms

When environmental winds are favourable, the updraft and downdraft of a storm become organized and twist around and reinforce each other. The result is a long-lived supercell storm. These storms are the most intense type of thunderstorm. In the central United States, supercells typically have a broad, intense updraft that enters from the southeast and brings moist surface air into the storm. The updraft rises, rotates counterclockwise, and exits to the east, forming an anvil. Updraft speeds in supercell storms can exceed 40 metres (130 feet) per second and are capable of suspending hailstones as large as grapefruit. Supercells can last two to six hours. They are the most likely storm to produce spectacular wind and hail damage as well as powerful tornadoes.

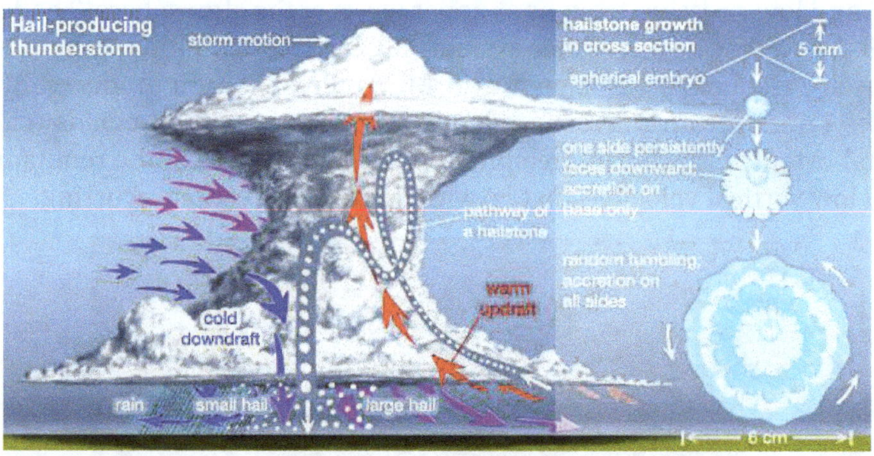

Hail-producing thunderstorm: (Left) A hailstone can travel through much of the height of the storm during its development and may make multiple vertical loops. (Right) Most hailstones are formed by accretion around a nucleus (spherical embryo). Peculiarly shaped hailstones are generally the product of multiple stones fusing together.

Physical Characteristics of Thunderstorms

Aircraft and radar measurements show that a single thunderstorm cell extends to an altitude of 8,000 to 10,000 metres (26,000 to 33,000 feet) and lasts about 30 minutes. An isolated storm usually contains several cells in different stages of evolution and lasts about an hour. A large storm can be many tens of kilometres in diameter with a top that extends to altitudes above 18 km (10 miles), and its duration can be many hours.

Updrafts and Downdrafts

The updrafts and downdrafts in isolated thunderstorms are typically between about 0.5 and 2.5 km (0.3 and 1.6 miles) in diameter at altitudes of 3 to 8 km (1.9 to 5 miles). The updraft diameter may occasionally exceed 4 km (2.5 miles). Closer to the ground, drafts tend to have a larger diameter and lower speeds than do drafts higher in the cloud. Updraft speeds typically peak in the range of 5 to 10 metres (16 to 33 feet) per second, and speeds exceeding 20 metres (66 feet) per second are common in the upper parts of large storms. Airplanes flying through large storms at altitudes of about 10,000 metres (33,000 feet) have measured updrafts exceeding 30 metres (98 feet) per second. The strongest updrafts occur in organized storms that are many tens of kilometres in diameter, and lines or zones of such storms can extend for hundreds of kilometres.

Downbursts

Sometimes thunderstorms will produce intense downdrafts that create damaging winds on the ground. These downdrafts are referred to as macrobursts or microbursts, depending on their size. A macroburst is more than 4 km (2.5 miles) in diameter and can produce winds as high as 60 metres per second, or 215 km per hour (200 feet per second, or 135 miles per hour). A microburst is smaller in dimension but produces winds as high as 75 metres per second, or 270 km per hour (250 feet per second, or 170 miles per hour) on the ground. When the parent storm forms in a wet, humid environment, the microburst will be accompanied by intense rainfall at the ground. If the storm forms in a dry environment, however, the precipitation may evaporate before it reaches the ground (such precipitation is referred to as virga), and the microburst will be dry.

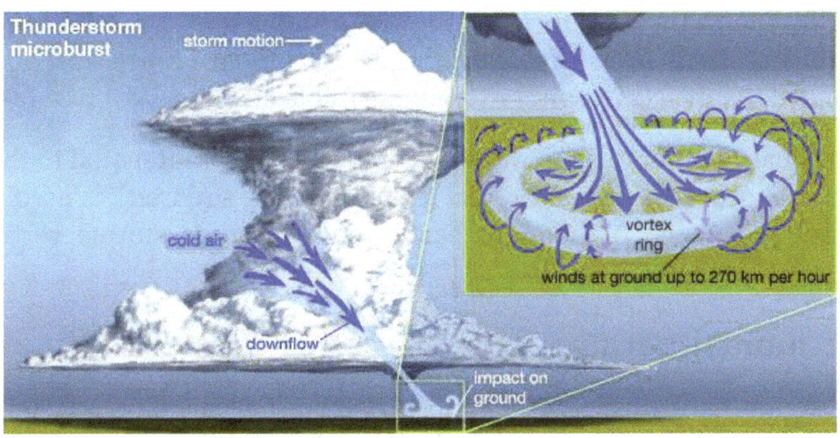

Thunderstorm microburst. (Left) The air that forms the microburst is initially "dammed" aloft by the strength of the storm's updraft then cascades downward in a high-velocity, narrow column (less than 4 km, or 2.5 miles, in diameter). (Right, inset) Microbursts are very dangerous to aircraft and can create great damage on the ground. In the absence of observers, microburst damage can often be distinguished from that of a tornado by the presence of a "starburst" pattern of destruction radiating from a central point.

Downbursts are a serious hazard to aircraft, especially during takeoffs and landings, because they produce large and abrupt changes in the wind speed and direction near the ground.

Vertical Extent

In general, an active cloud will rise until it loses its buoyancy. A loss of buoyancy is caused by precipitation loading when the water content of the cloud becomes heavy enough, or by the entrainment of cool, dry air, or by a combination of these processes. Growth can also be stopped by a capping inversion, that is, a region of the atmosphere where the air temperature decreases slowly, is constant, or increases with height.

Thunderstorms typically reach altitudes above 10,000 metres (33,000 feet) and sometimes more than 20,000 metres (66,000 feet). When the instability is high, the atmosphere moist, and winds favourable, thunderstorms can extend to the tropopause, that is, the boundary between the troposphere and the stratosphere. The tropopause is characterized by air temperatures that are nearly constant or increasing with height, and it is a region of great stability. Occasionally the momentum of an updraft carries it into the stratosphere, but after a short distance the air in the top of the updraft becomes cooler and heavier than the surrounding air, and the overshoot ceases. The height of the tropopause varies with both latitude and season. It ranges from about 10,000 to 15,000 metres (33,000 to 50,000 feet) and is higher near the Equator.

When a cumulonimbus cloud reaches a capping inversion or the tropopause, it spreads outward and forms the anvil cloud so characteristic of most thunderstorms. The winds at anvil altitudes typically carry cloud material downwind, and sometimes there are weak cells of convection embedded in the anvil.

Turbulence

An airplane flying through a thunderstorm is commonly buffeted upward and downward and from side to side by turbulent drafts in a storm. Atmospheric turbulence causes discomfort for the crew and passengers and also subjects the aircraft to undesirable stresses.

Turbulence can be quantified in various ways, but frequently a g unit, equal to the acceleration of gravity (9.8 metres per second squared, or 32.2 feet per second squared), is used. A gust of 1 g will cause severe aircraft turbulence. In the upper part of violent thunderstorms, vertical accelerations of about 3 g have been reported.

Movement of Thunderstorms

The motion of a thunderstorm across the land is determined primarily by the interactions of its updrafts and downdrafts with steering winds in the middle layers of the atmosphere in which the storm develops. The speed of isolated storms is typically about 20 km (12 miles) per hour, but some storms move much faster. In extreme circumstances, a supercell storm may move 65 to 80 km (about 40 to 50 miles) per hour. Most storms continually evolve and have new cells developing while old ones dissipate. When winds are light, an individual cell may move very little, less than two kilometres, during its lifetime; however, in a larger storm, new cells triggered by the outflow from downdrafts can give the appearance of rapid motion. In large, multicell storms, the new cells tend to form to the right of the steering winds in the Northern Hemisphere and to the left in the Southern Hemisphere.

Energy

The energy that drives thunderstorms comes primarily from the latent heat that is released when water vapour condenses to form cloud drops. For every gram of water that is condensed, about 600 calories of heat are released to the atmosphere. When water drops freeze in the upper parts of the cloud, another 80 calories per gram are released. The release of latent heat energy in an up-draft is converted, at least in part, to the kinetic energy of the air motions. A rough estimate of the total energy in a thunderstorm can be made from the total quantity of water that is precipitated by the cloud. In a typical case, this energy is about 107 kilowatt-hours, roughly equivalent of a 20-kiloton nuclear explosion (though it is released over a broader area and in a longer span of time). A large, multicell storm can easily be 10 to 100 times more energetic.

Weather under Thunderstorms

Downdrafts and Gust Fronts

Thunderstorm downdrafts originate at altitudes where the air temperature is cooler than at ground level, and they are kept cool even as they sink to warmer levels by the evaporation of water and melting of ice particles. Not only is the sinking air more dense than its surroundings, but it carries a horizontal momentum that is different from the surrounding air. If the descending air originated at a height of 10,000 metres (33,000 feet), for example, it might reach the ground with a horizontal velocity much higher than the wind at the ground. When such air hits the ground, it usually moves outward ahead of the storm at a higher speed than the storm itself. This is why an observer on the ground watching a thunderstorm approach can often feel a gust of cool air before the storm passes overhead. The outspreading downdraft air forms a pool some 500 to 2,000 metres (about 1,600 to 6,500 feet) deep, and often there is a distinct boundary between the cool air and the warm, humid air in which the storm developed. The passage of such a gust front is easily recognized as the wind speed increases and the air temperature suddenly drops. Over a five-minute period, a cooling of more than 5 °C (9 °F) is not unusual, and cooling twice as great is not unknown.

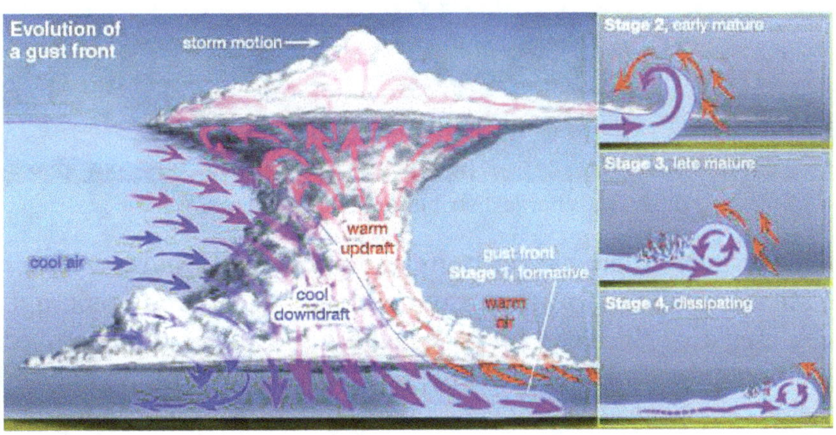

Evolution of a gust front (Left) During a thunderstorm a large column of cold air, originating high in the thundercloud, can descend rapidly to form a gust front. (Right, inset) Fed by the main downdraft, the gust front flows in a turbulent layer along the ground and can extend far from the main body of the storm. A gust front is often felt by observers as a sudden cool wind arriving well in advance of a storm.

Rainfall

In extreme circumstances, the gust front produced by a downburst may reach 50 metres (about 160 feet) per second or more and do extensive damage to property and vegetation. Severe winds occur most often when organized lines of thunderstorms develop in an environment where the middle-level winds are very strong. Under such conditions, people might think the winds were caused by a tornado. If a funnel cloud is not observed, the character of the wind damage can indicate the source. Tornadoes blow debris in a tight circular pattern, whereas the air from a thunderstorm outflow pushes it mostly in one direction.

By the time the cool air arrives, rain usually is reaching the surface. Sometimes all the raindrops evaporate while falling, and the result is a dry thunderstorm. At the other extreme, severe multiple-cell and supercell storms can produce torrential rain and hail and cause flash floods.

In small thunderstorms, peak five-minute rainfall rates can exceed 120 mm (4.7 inches) per hour, but most rainfalls are about one-tenth this amount. The average thunderstorm produces about 2,000 metric tons (220,000 short tons) of rain, but large storms can produce 10 times more rainfall. Large, organized storms that are associated with mesoscale convective systems can generate 10^{10} to 10^{12} kg of rainfall.

Thunderstorm Electrification

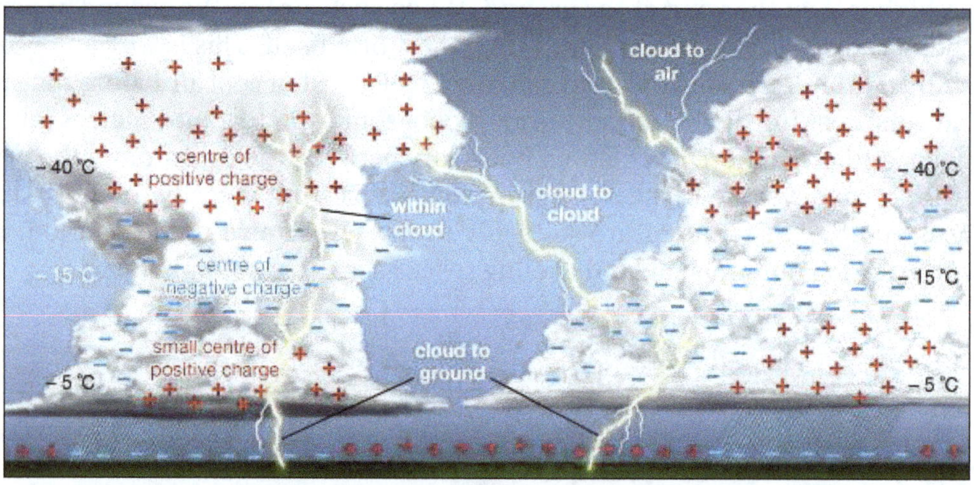

Electrical charge distribution in a thunderstorm.

Within a single thunderstorm, there are updrafts and downdrafts and a variety of cloud particles and precipitation. Measurements show that thunderclouds in different geographic locations tend to produce an excess negative charge at altitudes where the ambient air temperature is between about −5 and −15 °C (23 to 5 °F). Positive charge accumulates at both higher and lower altitudes. The result is a division of charge across space that creates a high electric field and the possibility of significant electrical activity.

When the electrical charges become sufficiently separated in a thundercloud, with some regions acquiring a negative charge and others a positive, a discharge of lightning becomes likely. About one-third of lightning flashes travel from the cloud to the ground; most of these originate in negatively charged regions of the cloud.

Many mechanisms have been proposed to explain the overall electrical structure of a thunderstorm, and cloud electrification is an active area of research. A leading hypothesis is that if the larger and heavier cloud particles charge preferentially with a negative polarity, and the smaller and lighter particles acquire a positive polarity, then the separation between positive and negative regions occurs simply because the larger particles fall faster than the lighter cloud constituents. Such a mechanism is generally consistent with laboratory studies that show electrical charging when soft hail, or graupel particles (porous amalgamations of frozen water droplets), collide with ice crystals in the presence of supercooled water droplets. The amount and polarity of the graupel charges depend on the ambient air temperature and on the liquid water content of the cloud, as well as on the ice crystal size, the velocity of the collision, and other factors. Other mechanisms of electrification are also possible.

Lightning Occurrence

Lightning flash striking a tree at a distance of 60 metres (200 feet) from the camera.

When the accumulated electric charges in a thunderstorm become sufficiently large, lightning discharges take place between opposite charge regions, between charged regions and the ground, or from a charged region to the neutral atmosphere. In a typical thunderstorm, roughly two-thirds of all discharges occur within the cloud, from cloud to cloud, or from cloud to air. The rest are between the cloud and ground.

Lightning discharge triggered by the presence of a tall tower atop Mount San Salvatore, near Lugano, Switzerland.

In recent years it has become clear that lightning can be artificially initiated, or triggered, in clouds that would not normally produce natural lightning discharges. Lightning can be triggered by a mountain or a tall structure when a thunderstorm is overhead and there is a high electric field in the vicinity or when an aircraft or large rocket flies into a high-field environment.

Global Lightning Distribution

Data from Earth-orbiting satellites show that, on average, about 80 percent of lightning flashes occur over land and 20 percent over the oceans. The frequency of lightning over land tends to peak in the mid-afternoon between 3:00 and 6:00 PM local time. Seasonal trends in the distribution of lightning are the result of temperature changes at the Earth's surface.

Tropical air masses commonly produce thunderstorms and lightning. Thunderstorm development requires moist, unstable air masses typical of those in tropical areas. In this region the Sun's rays are nearly vertical, allowing more energy to reach and warm the lowest layers of the atmosphere. Abundant moisture is added when the warm air moves over the ocean and becomes humidified by evaporation from the underlying water surface. Thunderstorm development is then initiated by upward movement of air, due to, for example, changes in air pressure or the topography of the land. The average number of days with audible thunder exceeds 100 per year over land areas within 10 degrees latitude north and south of the Equator. In some regions of equatorial Africa and South America there are more than 180 thunder days in an average year.

At higher latitudes, thunderstorm frequency depends on the character of the topography and how often moist, tropical air invades the region, which happens most often in the spring and summer. Maximum thunderstorm activity in the Northern and Southern Hemispheres is offset by approximately six months, with most Northern Hemisphere thunderstorms occurring between May and September and in the Southern Hemisphere between November and March.

Thunderstorms are a common feature of the summer monsoons in many parts of the world, especially southern Asia. As solar radiation warms the Indian subcontinent, an ocean-to-land air current is established and moist, unstable air from the Indian Ocean is carried inland. When this air is forced to rise by the steep slopes of the Himalayas, intense thunderstorms and rain showers are produced in great abundance.

In regions poleward of about 60 degrees latitude thunderstorms are rare to nonexistent. In these regions the air near the surface is cold and the atmosphere is generally stable. There are also few thunderstorms in regions that are dominated by semipermanent high-pressure centres, such as southern California. In these regions air from higher altitudes is descending and warming, which lowers the relative humidity and causes stable stratification of the lower atmosphere. As a result, thunderstorm development is inhibited.

Lightning Distribution in the United States

Every year, most of the United States experiences at least two cloud-to-ground strikes per square kilometre (about five per square mile). Most of the interior of the country east of the Rocky Mountains has four or more strikes per square kilometre (about 10 discharges per square mile). Summer thunderstorms are frequent in northern Mexico and the states of Arizona, New Mexico, and Colorado when warm, humid air is forced to rise by mountainous terrain.

Maximum flash densities are found along the Gulf Coast and Florida peninsula, where over a year's time, values exceeding 10 strikes per square kilometre (25 strikes per square mile) have been measured. More than 20 million cloud-to-ground flashes strike the United States annually, and lightning is clearly among the country's most severe weather hazards.

Cloud-to-Ground Lightning

Initial Stroke

A typical flash of cloud-to-ground lightning is initiated by electrical breakdown between the small positive charge region near the base of the cloud and the negative charge region in the middle of the cloud. The preliminary breakdown creates channels of air that have undergone partial ionization—the conversion of neutral atoms and molecules to electrically charged ones.

On timescales measured in fractions of a second, high-speed cameras can record luminous events in the flash. Initially, a faint luminous process descends in a downward-branching pattern in regular distinct steps, typically 30 metres (100 feet) in length, though they can range from 10 to 100 metres (33 to 330 feet). The time interval between steps ranges from 10 to 50 microseconds (millionths of a second). Carrying currents on the order of hundreds to thousands of amperes, the stepped leader propagates toward the ground at an average velocity of 1.5 × 105 metres per second, or about one two-thousandth the speed of light. It is called a stepped leader because of its downward-moving "stepped" pulses of luminosity. Diameter estimates for the stepped leader range from a few centimetres to a few metres. The current-carrying core has a diameter on the order of 1 or 2 cm (0.4 or 0.8 inch), and photographic measurements indicate that a corona sheath of electric charge with a diameter of 1 to 10 metres (3 to 33 feet) surrounds the core.

Return Stroke

As the stepped leader nears the ground, approximately five coulombs of charge have been deposited along the channel, inducing an opposite charge on the ground and increasing the electric field between the leader and the point to be struck. An upward discharge starts at the ground, church steeple, house, or other object, and rises to meet the stepped leader about 15 to 50 metres (50 to 160 feet) above the surface. At this moment of junction the cloud is short-circuited to the ground and a highly luminous return stroke of high current occurs. It is this return stroke, rather than the stepped leader, that is perceived as lightning because it is so much brighter and follows so quickly after the stepped leader. Portions of the stepped leader that have not reached the ground become the branches of the return stroke, and charge on the branches flows into the main channel. The five coulombs of charge typically deposited along the stepped leader flow to ground in a few hundred microseconds and produce peak currents that are usually on the order of 30,000 amperes but may range from a few thousand to over 200,000 amperes. Peak temperatures in the channel are on the order of 30,000 °C (50,000 °F), about five times hotter than the surface of the Sun. Because the junction process occurs near the ground, the time to peak current measured at the ground is typically only a few microseconds. As the leader charge avalanches toward the ground, the return stroke luminosity propagates toward the cloud base at an average speed of 5×10^7 to

2×10^8 metres per second, or approximately one-third the speed of light, and the high-current-carrying core expands to a diameter of a few centimetres. Laboratory experiments suggest that when pressure equilibrium is attained between the return stroke and the surrounding air, the channel approximates a high-current arc characterized by a current density of 1,000 amperes per square centimetre.

Cloud-to-ground lightning discharge showing a bright main channel and secondary branches.

Subsequent Return Strokes

In the rapid passage from ground to cloud, the luminous return stroke is observed to pause at points where large branches join the main channel, and the channel is observed to brighten as charge from the branch flows into the channel. The stroke then continues its upward propagation, reaching the level of the atmosphere where the temperature is 0 °C (typically at an altitude of 5 km [3 miles] above sea level) in approximately 100 microseconds; the downward-propagating stepped leader traverses the same distance in about 30 milliseconds (thousandths of a second). There is then a pause for tens of milliseconds, and the channel cools to a few thousand degrees Celsius. If a second stroke occurs, it begins with the appearance of a dart of light, perhaps 30 to 50 metres (100 to 160 feet) in length, propagating down the channel of the previous return stroke. The dart leader moves downward at a speed of 2×10^6 metres per second (about one one-hundredth the speed of light) and carries a current of the order of 1,000 amperes toward the ground. Once again, when the leader effectively short-circuits a charge centre in the cloud to the ground, another return stroke occurs. After the first stroke, the dart leader may follow the lightning channel only partway before taking a new path to the ground. This gives rise to the common forked appearance of lightning as it strikes the ground.

This sequence of dart leader-return stroke typically occurs three to four times, although a flash to the ground that had 26 strokes and lasted two seconds has been reported. When a flash does have more than one stroke, the subsequent return strokes draw charge from different regions of the parent thunderstorm. Multiple strokes of lightning appear to flicker because the human eye is just capable of resolving the time interval between them.

Dissipation of Energy

During the return-stroke stage, approximately 10^5 joules of energy per metre are dissipated within the lightning channel. This energy is divided among the dissociation, ionization, excitation, and kinetic energy of the particles, the energy of expansion of the channel, and radiation. Spectroscopic measurements reveal that the air molecules, principally those of nitrogen, oxygen, and water, are split into their respective atoms and that on the average one electron is removed from each atom. The conversion from neutral air molecules to a completely ionized plasma occurs in a few microseconds.

Thunder

When the stroke plasma is created, its temperature is at least 30,000 °C (50,000 °F), and the pressure is greater than 1,000 kilopascals (10 atmospheres). The channel pressure greatly exceeds the ambient (surrounding) pressure, and the return-stroke channel expands at a supersonic rate. The resultant shock wave decays rapidly with distance and is eventually heard as thunder once it slows to the speed of sound. Because it is estimated that only 1 percent of the input energy is stored in the particles and less than 1 percent is emitted as radiation in the visible and infrared region (4,000 to 11,000 angstroms [Å], where Å = 10–10 metre), it is probable that most of the energy dissipated goes into the energy of channel expansion, a process requiring no more than 10 to 20 microseconds.

Since light travels at about 300,000 km (186,000 miles) per second and the speed of sound is only about 0.33 km (0.2 mile) per second, the light from a discharge will always be seen before the sound arrives at an observer. The time delay between the bright flash of light and the arrival of the associated thunder can often be used to estimate the distance to a discharge. Every three seconds correspond to one kilometre, and every five seconds correspond to one mile.

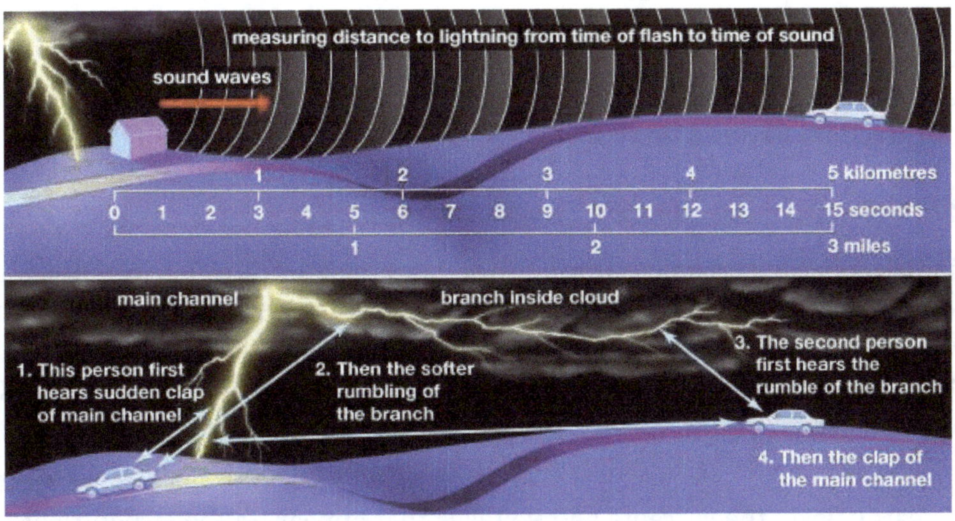

(Top) As shown in the chart, the elapsed time between seeing a flash of lightning and hearing the thunder is roughly three seconds for each kilometre, or five seconds for each mile. (Bottom) An observer's relative distance from the main lightning channel and its secondary branches determines whether thunder is heard to start with a sudden clap or a softer rumbling.

The total thunder waveform comes from the entire lightning channel and includes the effects of channel branching and tortuosity, sound propagation in the atmosphere, and acoustic reflections from the local topography. The result is a series of sounds that are variously described as peaks, claps, rolls, and rumbles. At distances of a few hundred metres, thunder begins with a sudden clap followed by a long rumble; at larger distances, it begins with a rumble.

Triggered Lightning

A small percentage of discharges between the cloud and ground are actually initiated at the ground and propagate upward to a charged region in the cloud. These discharges often are initiated (or triggered) by tall structures or by towers on hilltops. The upward branching of such discharges makes them visually distinguishable from their "right-side-up" counterparts, giving the impression of a cloud-to-ground lightning flash that is upside down.

Cloud-to-Cloud and Intracloud Lightning

True cloud-to-cloud lightning is rare because most lightning flashes occur within a cloud. The first lightning flash in a thunderstorm is typically an intracloud discharge. When an intracloud discharge occurs, the cloud becomes luminous for approximately 0.2 to 0.5 second. The discharge is initiated by a leader that propagates between regions of opposite charge (or from a charged region to the neutral atmosphere). Luminosity is more or less continuous and has several pulses of higher luminosity of one-millisecond duration superimposed upon it. This situation suggests minor return strokes as the leader contacts pockets of opposite charge, but the similarity ends there. The total amount of the charge transfer is generally similar to the amount involved in a ground discharge: 10 coulombs, with a range from 0.3 to 100 coulombs. The mean velocity of propagation of intracloud lightning ranges from 10^4 to 10^7 metres per second. Electric currents associated with the luminous brightening are probably in the range of 1,000 to 4,000 amperes. Strikes to aircraft exhibit peak currents of only a few thousand amperes, about an order of magnitude less than currents in ground flashes—though sometimes the peak currents are large. Rise times to peak currents in cloud flashes are generally slower than those in return strokes. The amount of energy dissipated by intracloud flashes is unknown.

Lightning Damage

Most lightning strikes cause damage through the large current flowing in the return stroke or through the heat that is generated by this and the continuing current. The precise mechanisms whereby lightning currents cause damage are not completely understood, however. If lightning strikes a person, the stroke current can damage the central nervous system, heart, lungs, and other vital organs.

When a building or power line is struck by lightning or is exposed to the intense electromagnetic fields from a nearby flash, the currents and voltages that appear on the structure are determined both by the currents and fields in the discharge and by the electrical response of the object and its grounding system. For instance, if a lightning surge enters an unprotected residence by way of an electric power line, the voltages may be large enough to cause sparks in the house wiring or appliances. When such flashovers occur, they may short-circuit the alternating current power system, and the resulting power arc may start a fire. In such instances, the lightning does not start the fire directly, but it does cause a power fault (short circuit), and then the power currents do the damage. In the case of metals, large currents heat the surface at the air-arc interface and the interior by electron collisions with the metal lattice. If this heat is also great enough, the metal will melt or evaporate.

At least three properties of the return-stroke current can cause damage; these are the peak current, the maximum rate of change of the initial current, and the total amount of charge transferred. For objects that have a resistive impedance, such as a ground rod or a long power line, the peak voltage during a strike is proportional to the peak current produced of the lightning stroke and the resistivity of the struck object. For example, if a 100,000 ampere peak current flows into a 10-ohm grounding system, 1 million volts will be produced. A common hazard associated with the large voltages produced by lightning strikes is the re-direction of some of the energy (that is, a flash-over) from the original target to an adjacent object. Such secondary discharges, or side-flashes, often cause damage comparable to that of a direct strike, and they are one of the main hazards of standing under or near an isolated tree (or any other tall object) during a thunderstorm. Such large voltages frequently cause secondary discharges or side-flashes to radiate outward from the object that is struck to another object nearby. One form of a side-flash can even occur in the ground near the point of lightning attachment.

For objects that have an inductive electrical impedance, such as the wires in a home electrical system, the peak voltage will be proportional to the maximum rate of change of the lightning current and the inductance of the object. For example, one metre of straight copper wire has a self-inductance on the order of one microhenry. The peak rate of change in the lightning current in a return stroke is on the order of 100,000 amperes per microsecond; therefore, about 100,000 volts will appear across this length of conductor for the duration of the change, typically 100 nanoseconds (billionths of a second).

The heating and subsequent burn-through of metal sheets, as on a metal roof or tank, are to a first approximation proportional to the total charge injected into the metal at the air-arc interface. Generally, large charge transfers are produced by long-duration continuing currents that are in the range of 100 to 1,000 amperes, rather than by the peak currents, which have a relatively short duration. The heat produced by long continuing currents is frequently the cause of forest fires. A typical cloud-to-ground flash transfers 20 to 30 coulombs of charge to the ground, and extreme flashes transfer hundreds and occasionally thousands of coulombs.

Lightning Protection

The best personal protection against lightning is to be alert to the presence of a hazard and then to take common-sense precautions, such as staying inside a house or building or inside an automobile, where one is surrounded by (but not in contact with) metal. People are advised to stay away from outside doors and windows and not to be in contact with any electrical appliances, such as a telephone, or anything connected to the plumbing system. If caught outdoors, people are advised to avoid isolated trees or other objects that are preferred targets and to keep low so as to minimize both height and contact with the ground (that is, crouch but do not lie down). Swimming pools are not safe during a lightning storm because water is a good conductor of electricity, and hence being in the pool effectively greatly multiplies the area of one's "ground" contact

The frequency with which lightning will directly strike a building in a particular region can be estimated from the building's size and the average number of strikes that occur in the region. If a building is struck whenever a stepped leader comes within 10 metres (33 feet) of the exterior of the building, then a building that is 12 metres (39 feet) wide and 16 metres (52 feet) long (an area of 192 square metres, or about 2,000 square feet) will have an effective strike zone of 32 metres by 36 metres (an area of 1,152 square metres, or 12,400 square feet). In a region where an average of three cloud-to-ground lightning strikes occur per square kilometre annually, such a building will experience an average of 0.0035 direct strike per year, or one strike about every 290 years (1,152 square metres × 3 flashes per square kilometre × 10–6 metres per square kilometre). In a region where there is an annual average of five strikes per square kilometre, the same building will experience an average of 0.0058 direct strike per year, or one strike about every 174 years. These calculations indicate that, for the second example, an average of one of every 174 buildings of similar size will be directly struck by lightning in that region each year.

Structures may be protected from lightning by either channeling the current along the outside of the building and into the ground or by shielding the building against damage from transient currents and voltages caused by a strike. Many buildings constrain the path of lightning currents and voltages through use of lightning rods, or air terminals, and conductors that route the current down into a grounding system. When a lightning leader comes near the building, the lightning rod initiates a discharge that travels upward and connects with it, thus controlling the point of attachment of lightning to the building. A lightning rod functions only when a lightning strike in the immediate vicinity is already immanent and so does not attract significantly more lighting to the building. The down conductors and grounding system function to guide the current into the ground while minimizing damage to the structure. To minimize side-flashes, the grounding resistance should be kept as low as possible, and the geometry should be arranged so as to minimize surface breakdown. Overhead wires and grounded vertical cones may also be used to provide a cone-shaped area of lightning protection. Such systems are most efficient when their height is 30 metres (98 feet) or less.

Protection of the contents of a structure can be enhanced by using lightning arresters to reduce any transient currents and voltages that might be caused by the discharge and that might propagate into the structure as traveling waves on any electric power or telephone wires exposed to the outside environment. The most effective protection for complex structures is provided by topological shielding. This form of protection reduces amounts of voltage and power at each level of a system of successive nested shields. The partial metallic shields are isolated, and the inside surface of each is grounded to the outside surface of the next. Power surges along wires coming into the structure are deflected by arrestors, or transient protectors, to the outside surface of each shield as they travel through the series, and are thus incrementally attenuated.

JET STREAM

Jet streams are relatively narrow bands of strong wind in the upper levels of the atmosphere. The winds blow from west to east in jet streams but the flow often shifts to the north and south. Jet streams follow the boundaries between hot and cold air.

Since these hot and cold air boundaries are most pronounced in winter, jet streams are the strongest for both the northern and southern hemisphere winters.

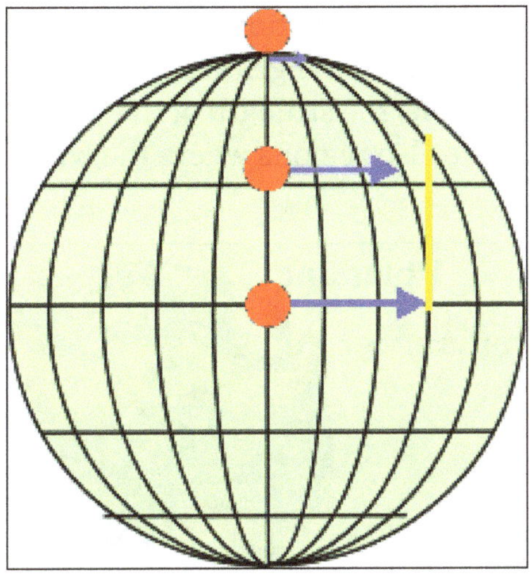

Earth's rotation the effects the west to east direction of the jet stream.

We saw that the earth's rotation divided this circulation into three cells. The earth's rotation is responsible for the jet stream as well.

The motion of the air is not directly north and south but is affected by the momentum the air has as it moves away from the equator. The reason has to do with momentum and how fast a location on or above the Earth moves relative to the Earth's axis.

Your speed relative to the Earth's axis depends on your location. Someone standing on the equator is moving much faster than someone standing on a 45° latitude line. In the graphic (above right) the person at the position on the equator arrives at the yellow line sooner than the other two.

Someone standing on a pole is not moving at all (except that he or she would be slowly spinning). The speed of the rotation is great enough to cause you to weigh one pound less at the equator than you would at the north or south pole.

The momentum the air has as it travels around the earth is conserved, which means as the air that's over the equator starts moving toward one of the poles, it keeps its eastward motion constant. The Earth below the air, however, moves slower as that air travels toward the poles. The result is that the air moves faster and faster in an easterly direction (relative to the Earth's surface below) the farther it moves from the equator.

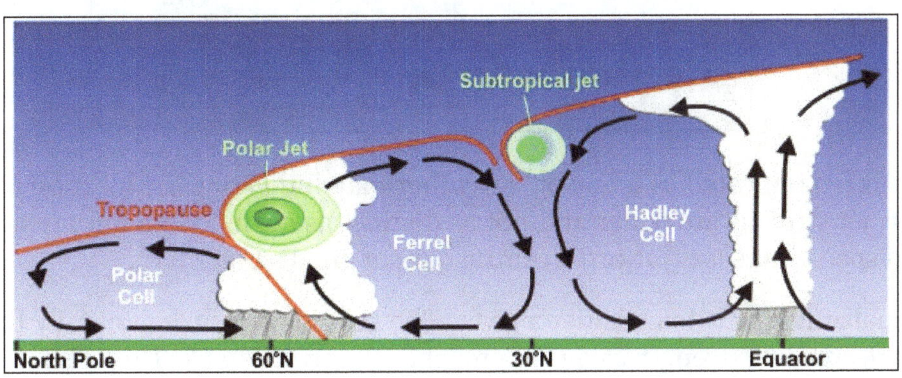

North hemisphere cross section showing jet streams and tropopause elevations.

In addition, with the three-cell circulations mentioned previously, the regions around 30° N/S and 50°-60° N/S are areas where temperature changes are the greatest. As the difference in temperature increases between the two locations the strength of the wind increases. Therefore, the regions around 30° N/S and 50°-60° N/S are also regions where the wind, in the upper atmosphere, is the strongest.

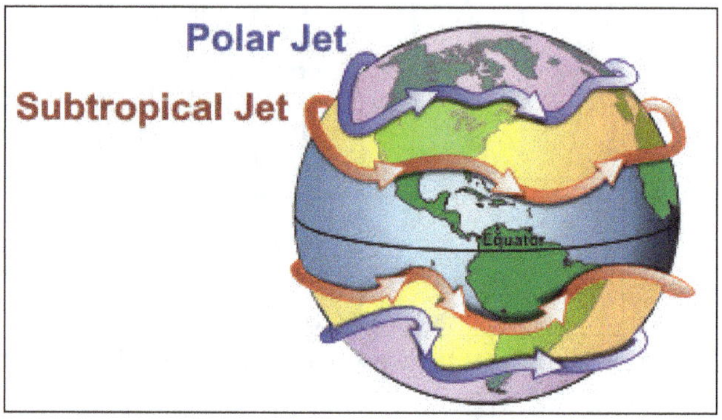

The 50°-60° N/S region is where the polar jet located with the subtropical jet located around 30°N. Jet streams vary in height of four to eight miles and can reach speeds of more than 275 mph (239 kts / 442 km/h).

The actual appearance of jet streams result from the complex interaction between many variables - such as the location of high and low pressure systems, warm and cold air, and seasonal changes. They meander around the globe, dipping and rising in altitude/latitude, splitting at times and forming eddies, and even disappearing altogether to appear somewhere else.

Jet streams also "follow the sun" in that as the sun's elevation increases each day in the spring, the average latitude of the jet stream shifts poleward. (By Summer in the Northern Hemisphere, it is typically found near the U.S. Canadian border.) As Autumn approaches and the sun's elevation decreases, the jet stream's average latitude moves toward the equator.

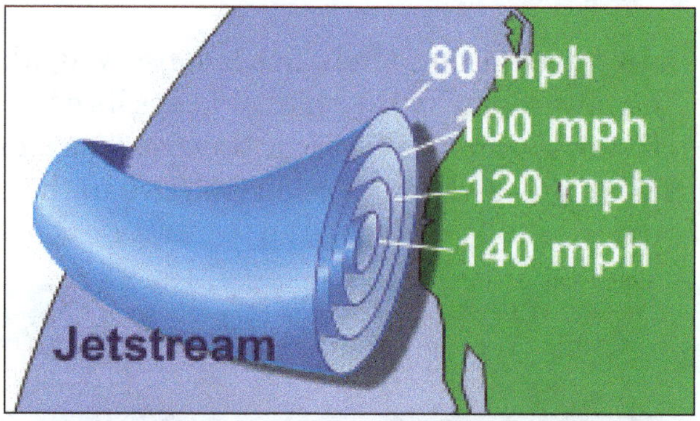

Jet streams are often indicated by a line on a weather map indicating the location of the strongest wind. However, jet streams are wider and not as distinct as a single line but are actually regions where the wind speed increases toward a central core of greatest strength.

One way of visualizing this is to consider a river. The river's current is generally the strongest in the center with decreasing strength as one approaches the river's bank. Therefore, it is said that jet streams are "rivers of air".

References

- Greenflash, atmospheric: phenomena.org, Retrieved 16 June, 2019
- Feldstein, y. I. (2011). "a quarter century with the auroral oval". Eos. 67 (40): 761. Bibcode:1986eostr..67..761f. Doi:10.1029/e0067i040p00761-02
- Beltofvenus atmospheric: phenomena.org, Retrieved 21 August, 2019
- Clark, stuart (2007). "astronomical fire: richard carrington and the solar flare of 1859". Endeavour. 31 (3): 104–109. Doi:10.1016/j.endeavour.2007.07.004
- 22halo, atmospheric: phenomena.org, Retrieved 1 April, 2019
- Tornado, science: britannica.com, Retrieved 14 June, 2019
- Jet, jetstream: weather.gov, Retrieved 21 July, 2019
- Tropical-cyclone, science: britannica.com, Retrieved 28 May, 2019
- Ryan n. Maue (2004-12-07). "chapter 3: cyclone paradigms and extratropical transition conceptualizations". Archived from the original on 2008-05-10. Retrieved 2008-06-15

5

Key Concepts in Atmospheric Science

Atmospheric temperature, atmospheric refraction, atmospheric density, atmospheric circulation, atmospheric electricity, diffuse sky radiation and atmospheric pressure are some of the key concepts of atmospheric science. The chapter closely examines these key concepts of atmospheric science to provide an extensive understanding of the subject.

ATMOSPHERIC TEMPERATURE

The temperature of Earth's atmosphere varies with the distance from the equator (latitude) and height above the surface (altitude). It also changes with time, varying from season to season, and from day to night, as well as irregularly due to passing weather systems. If local variations are averaged out on a global basis, however, a pattern of global average temperatures emerges. Vertically, the atmosphere is divided into four layers: the troposphere, the stratosphere, the mesosphere, and the thermosphere.

The Vertical Temperature Profile

Averaging atmospheric temperatures over all latitudes and across an entire year gives us the average vertical temperature profile that is known as a standard atmosphere. The average vertical temperature profile suggests four distinct layers. In the first layer, known as the troposphere, average atmospheric temperature drops steadily from its value at the surface, about 290K (63 °F; 17 °C) and reaches of minimum of around 220K (−64 °F;−53 °C) at an altitude of about 6.2 mi (10 km). This level, known as the tropo-pause, is just above the cruising altitude of commercial jet aircraft. The decrease in temperature with height, called the lapse rate, is nearly steady throughout the troposphere at 43.7 °F (6.5 °C) per 0.6 mi (1 km). At the tropopause, the lapse rate abruptly decreases. Atmospheric temperature is nearly constant over the next 12 mi (20 km), then begins to rise with increasing altitude up to about 31 mi (50 km). This region of increasing temperatures is the stratosphere. At the top of the layer, called the stratopause, temperatures are nearly as warm as the surface values. Between about 31–50 mi (50–80 km) lies the mesosphere, where atmospheric temperature resumes its decrease with altitude and reaches a minimum of 180K (−136 °F; −93 °C) at the top of the layer (the mesopause), around 50 mi (80 km). Above the mesopause is the thermosphere that, as its name implies, is a zone of high gas temperatures. In the very high thermosphere (about 311 mi (500 km) above Earth's surface) gas temperatures can reach from

500–2,000K (441–3, 141 °F; 227–1, 727 °C). Temperature is a measure of the energy of the gas molecules' motion. Although they have high energy, the molecules in the thermosphere are present in very low numbers, less than one millionth of the amount present on average at Earth's surface.

Atmospheric temperature can also be plotted as a function of both latitude and altitude. Show such plots, with latitude as the x coordinate and altitude as the y.

The Sun's Role in Atmospheric Temperature

Most solar radiation is emitted as visible light, with smaller portions at shorter wavelengths (ultraviolet radiation) and longer wavelengths (infrared radiation, or heat). Little of the visible light is absorbed by the atmosphere (although some is reflected back into space by clouds), so most of this energy is absorbed by Earth's surface. The Earth is warmed in the process and radiates heat (infrared radiation) back upward. This warms the atmosphere, and, just as one will be warmer when standing closer to a fire, the layers of air closest to the surface are the warmest.

According to this explanation, the temperature should continually decrease with altitude. However, shows that temperature increaseS with altitude in the stratosphere. The stratosphere contains nearly all the atmosphere's ozone. Ozone (O_3) and molecular oxygen (O_2) absorb most of the sun's short wavelength ultraviolet radiation. In the process they are broken apart and reform continuously. The net result is that the ozone molecules transform the ultraviolet radiation to heat energy, heating up the layer and causing the increasing temperature profile observed in the stratosphere.

The mesosphere resumes the temperature decrease with altitude. The thermosphere, however, is subject to very high energy, short wavelength ultraviolet and x-ray solar radiation. As the atoms or molecules present at this level absorb some of this energy, they are ionized

(Have an electron removed) or dissociated (molecules are split into their component atoms). The gas layer is strongly heated by this energy bombardment, especially during periods when the sun is emitting elevated amounts of short wavelength radiation.

The Greenhouse Effect

Solar energy is not the only determinant of atmospheric temperature. Earth's surface, after absorbing solar radiation in the visible region, emits infrared radiation back to space. Several atmospheric gases absorb this heat radiation and re-radiate it in all directions, including back toward the surface. These so-called greenhouse gases thus trap infrared radiation within the atmosphere, raising its temperature. Important greenhouse gases include water vapor (H_2O), carbon dioxide (CO_2), and methane (CH_4). It is estimated that Earth's surface temperature would average about 32 °C (90 °F) cooler in the absence of greenhouse gases. Because this temperature is well below the freezing point of water, the planet would be much less hos-pitable to life in the absence of the greenhouse effect.

While greenhouse gases are essential to life on the planet, more is not necessarily better. Since the beginning of the industrial revolution in the mid-nineteenth century, humans have released increasing amounts of carbon dioxide to the atmosphere by burning fossil fuels. The level of carbon dioxide measured in the remote atmosphere has shown a continuous increase since record keeping began in 1958. If this increase translates into a corresponding rise in atmospheric temperature, the results might include melting polar ice caps and swelling seas, resulting in coastal cities being covered by the ocean; shifts in climate perhaps leading to extinctions; and unpredictable changes in wind and weather patterns, posing significant challenges for agriculture. Predicting the changes that increased levels of greenhouse gases may bring is complicated. The interaction of the atmosphere, the oceans, the continents, and the ice caps is not completely understood. While it is known that some of the emitted carbon dioxide is absorbed by the oceans and eventually deposited as carbonate rock (such as limestone), it is not known if this is a steady process or if it can keep pace with current levels of carbon dioxide production.

ATMOSPHERIC PRESSURE

Atmospheric pressure, sometimes also called barometric pressure (after the sensor), is the pressure within the atmosphere of Earth (or that of another planet). The standard atmosphere (symbol: atm) is a unit of pressure defined as 1013.25 mbar (1013.25 hPa), equivalent to 760 mm Hg (torr), 29.9212 inches Hg, or 14.696 psi. The atm unit is roughly equivalent to the mean sea-level atmospheric pressure on Earth, that is, the Earth's atmospheric pressure at sea level is approximately 1 atm.

In most circumstances atmospheric pressure is closely approximated by the hydrostatic pressure caused by the weight of air above the measurement point. As elevation increases, there is less overlying atmospheric mass, so that atmospheric pressure decreases with increasing elevation. Pressure measures force per unit area, with SI units of Pascals (1 pascal = 1 newton per square metre, 1 N/m^2). On average, a column of air with a cross-sectional area of 1 square centimetre (cm^2), measured from mean (average) sea level to the top of Earth›s atmosphere, has a mass of about 1.03 kilogram and exerts a force or "weight" of about 10.1 newtons, resulting in a pressure of 10.1 N/cm^2 or 101 kN/m^2 (101 kilopascals, kPa). A column of air with a cross-sectional area of 1 in^2 would have a weight of about 14.7 lb_f, resulting in a pressure of 14.7 lb_f/in^2.

Mechanism

Atmospheric pressure is caused by the gravitational attraction of the planet on the atmospheric gases above the surface, and is a function of the mass of the planet, the radius of the surface, and the amount and composition of the gases and their vertical distribution in the atmosphere. It is modified by the planetary rotation and local effects such as wind velocity, density variations due to temperature and variations in composition.

Mean Sea-level Pressure

The *mean sea-level pressure* (MSLP) is the average atmospheric pressure at mean sea level. This is the atmospheric pressure normally given in weather reports on radio, television, and newspapers or on the Internet. When barometers in the home are set to match the local weather reports, they measure pressure adjusted to sea level, not the actual local atmospheric pressure.

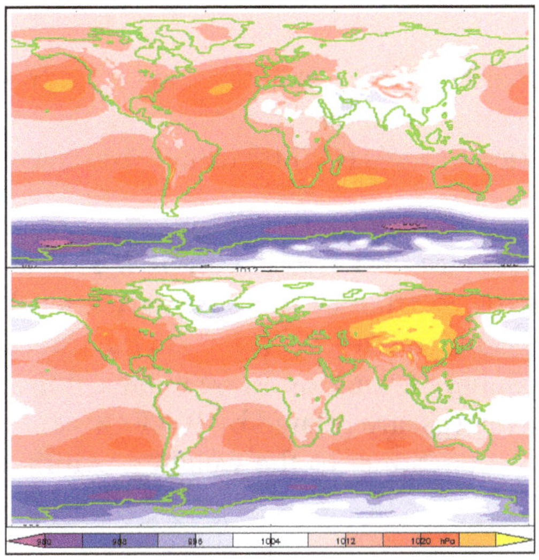

15-year average mean sea-level pressure for June, July, and August (top) and December, January, and February (bottom). ERA-15 re-analysis.

Kollsman-type barometric aircraft altimeter (as used in North America) displaying an altitude of 80 ft (24 m).

The altimeter setting in aviation is an atmospheric pressure adjustment.

Average sea-level pressure is 1013.25 mbar (101.325 kPa; 29.921 inHg; 760.00 mmHg). In aviation, weather reports (METAR), QNH is transmitted around the world in millibars or hectopascals (1 hectopascal = 1 millibar), except in the United States, Canada, and Colombia where it is reported in inches of mercury (to two decimal places). The United States and Canada also report sea-level pressure SLP, which is adjusted to sea level by a different method, in the remarks section, not in the internationally transmitted part of the code, in hectopascals or millibars. However, in Canada›s public weather reports, sea level pressure is instead reported in kilopascals.

In the US weather code remarks, three digits are all that are transmitted; decimal points and the one or two most significant digits are omitted: 1013.2 mbar (101.32 kPa) is transmitted as 132; 1000.0 mbar (100.00 kPa) is transmitted as 000; 998.7 mbar is transmitted as 987; etc. The highest *sea-level pressure* on Earth occurs in Siberia, where the Siberian High often attains a *sea-level pressure* above 1050 mbar (105 kPa; 31 inHg), with record highs close to 1085 mbar (108.5 kPa; 32.0 inHg). The lowest measurable *sea-level pressure* is found at the centers of tropical cyclones and tornadoes, with a record low of 870 mbar (87 kPa; 26 inHg).

Surface Pressure

Surface pressure is the atmospheric pressure at a location on Earth's surface (terrain and oceans). It is directly proportional to the mass of air over that location. For numerical reasons, atmospheric models such as general circulation models (GCMs) usually predict the nondimensional *logarithm of surface pressure*.

The average value of surface pressure on Earth is 985 hPa. This is in contrast to mean sea-level pressure, which involves the extrapolation of pressure to sea-level for locations above or below sea-level. The average pressure at mean sea-level (MSL) in the International Standard Atmosphere (ISA) is 1013.25 hPa, or 1 atmosphere (Atm), or 29.92 inches of mercury. Pressure (P), mass (m), and the acceleration due to gravity (g), are related by $P = F/A = (m*g)/A$, where A is surface area. Atmospheric pressure is thus proportional to the weight per unit area of the atmospheric mass above that location.

Altitude Variation

A very local storm above Snæfellsjökull, showing clouds formed on the mountain by orographic lift.

Pressure on Earth varies with the altitude of the surface; so air pressure on mountains is usually lower than air pressure at sea level. Pressure varies smoothly from the Earth's surface to the top of the mesosphere. Although the pressure changes with the weather, NASA has averaged the conditions for all parts of the earth year-round. As altitude increases, atmospheric pressure decreases. One can calculate the atmospheric pressure at a given altitude. Temperature and humidity also affect the atmospheric pressure, and it is necessary to know these to compute an accurate figure. The graph at right was developed for a temperature of 15 °C and a relative humidity of 0%.

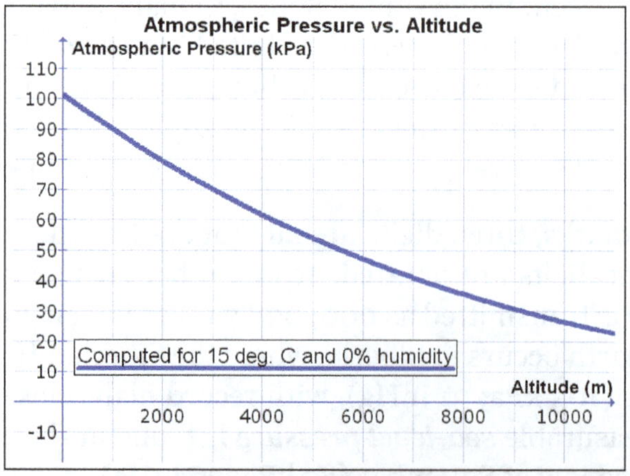

Variation in atmospheric pressure with altitude, computed for 15 °C and 0% relative humidity.

This plastic bottle was sealed at approximately 14,000 feet (4,300 m) altitude, and was crushed by the increase in atmospheric pressure, recorded at 9,000 feet (2,700 m) and 1,000 feet (300 m), as it was brought down towards sea level.

At low altitudes above sea level, the pressure decreases by about 1.2 kPa for every 100 metres. For higher altitudes within the troposphere, the following equation (the barometric formula) relates atmospheric pressure *p* to altitude *h*, where the constant parameters are as described below:

$$p = p_0 \cdot \left(1 - \frac{L \cdot h}{T_0}\right)^{\frac{g \cdot M}{R_0 \cdot L}}$$

$$= p_0 \cdot \left(1 - \frac{g \cdot h}{c_p \cdot T_0}\right)^{\frac{c_p \cdot M}{R_0}} \approx p_0 \cdot \exp\left(-\frac{g \cdot h \cdot M}{T_0 \cdot R_0}\right)$$

Parameter	Description	Value
p_o	Sea level standard atmospheric pressure	101325 Pa
L	Temperature lapse rate, = g/c_p for dry air	~ 0.00976 K/m
c_p	Constant-pressure specific heat	1004.68506 J/(kg·K)
T_o	Sea level standard temperature	288.16 K
g	Earth-surface gravitational acceleration	9.80665 m/s²
M	Molar mass of dry air	0.02896968 kg/mol
R_o	Universal gas constant	8.314462618 J/(mol·K)

Local Variation

Atmospheric pressure varies widely on Earth, and these changes are important in studying weather and climate. Atmospheric pressure shows a diurnal or semidiurnal (twice-daily) cycle caused by global atmospheric tides. This effect is strongest in tropical zones, with an amplitude of a few millibars, and almost zero in polar areas. These variations have two superimposed cycles, a circadian (24 h) cycle and semi-circadian (12 h) cycle.

Hurricane Wilma on 19 October 2005; 882 hPa (12.79 psi) in the storm's eye.

Records

The highest adjusted-to-sea level barometric pressure ever recorded on Earth (above 750 meters) was 1084.8 hPa (32.03 inHg) measured in Tosontsengel, Mongolia on 19 December 2001. The highest adjusted-to-sea level barometric pressure ever recorded (below 750 meters) was at Agata in Evenk Autonomous Okrug, Russia (66°53' N, 93°28' E, elevation: 261 m, 856 ft) on 31 December 1968 of 1083.8 hPa (32.005 inHg). The discrimination is due to the problematic assumptions (assuming a standard lapse rate) associated with reduction of sea level from high elevations.

The Dead Sea, the lowest place on Earth at 430 metres (1,410 ft) below sea level, has a correspondingly high typical atmospheric pressure of 1065 hPa.

The lowest non-tornadic atmospheric pressure ever measured was 870 hPa (0.858 atm; 25.69 inHg), set on 12 October 1979, during Typhoon Tip in the western Pacific Ocean. The measurement was based on an instrumental observation made from a reconnaissance aircraft.

Measurement based on Depth of Water

One atmosphere (101.325 kPa or 14.7 psi) is also the pressure caused by the weight of a column of fresh water of approximately 10.3 m (33.8 ft). Thus, a diver 10.3 m underwater experiences a pressure of about 2 atmospheres (1 atm of air plus 1 atm of water). Conversely, 10.3 m is the maximum height to which water can be raised using suction under standard atmospheric conditions.

Low pressures such as natural gas lines are sometimes specified in inches of water, typically written as w.c. (water column) gauge or w.g. (inches water gauge). A typical gas-using residential appliance in the US is rated for a maximum of 14 w.g., which is approximately 1048.37 hPa. Similar metric units with a wide variety of names and notation based on millimetres, centimetres or metres are now less commonly used.

Measurement and Maps

An important application of the knowledge that atmospheric pressure varies directly with altitude was in determining the height of hills and mountains thanks to the availability of reliable pressure measurement devices. While in 1774, Maskelyne was confirming Newton's theory of gravitation at and on Schiehallion in Scotland (using plumb bob deviation to show the effect of gravity) and accurately measure elevation, William Roy using barometric pressure was able to confirm his height determinations, the agreement being to within one meter (3.28 feet). This method became and continues to be useful for survey work and map making. This early application of science gave people insight into how science could easily be put to practical use.

ATMOSPHERIC REFRACTION

Atmospheric refraction is the deviation of light or other electromagnetic wave from a straight line as it passes through the atmosphere due to the variation in air density as a function of height. This refraction is due to the velocity of light through air, decreasing (the refractive index increases) with increased density. Atmospheric refraction near the ground produces mirages. Such refraction can also raise or lower, or stretch or shorten, the images of distant objects without involving mirages. Turbulentair can make distant objects appear to twinkle or shimmer. The term also applies to the refraction of sound. Atmospheric refraction is considered in measuring the position of both celestial and terrestrial objects.

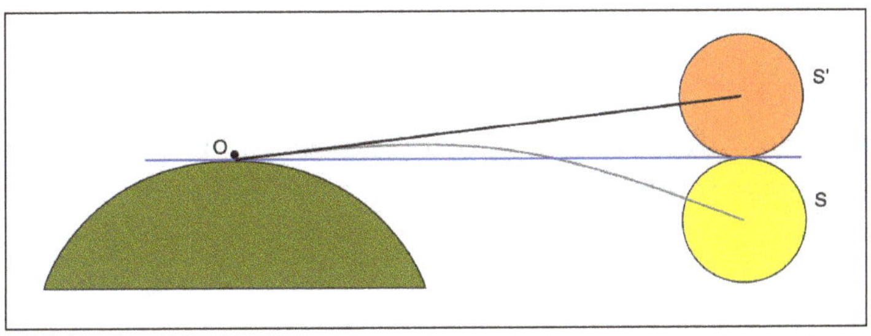

Diagram showing displacement of the Sun's image at sunrise and sunset.

Astronomical or celestial refraction causes astronomical objects to appear higher above the horizon than they actually are. Terrestrial refraction usually causes terrestrial objects to appear higher than they actually are, although in the afternoon when the air near the ground is heated, the rays can curve upward making objects appear lower than they actually are.

Refraction not only affects visible light rays, but all electromagnetic radiation, although in varying degrees. For example, in the visible spectrum, blue is more affected than red. This may cause astronomical objects to appear dispersed into a spectrum in high-resolution images.

The atmosphere refracts the image of a waxing crescent Moon as it sets into the horizon.

Whenever possible, astronomers will schedule their observations around the times of culmination, when celestial objects are highest in the sky. Likewise, sailors will not shoot a star below 20° above the horizon. If observations of objects near the horizon cannot be avoided, it is possible to equip an optical telescope with control systems to compensate for the shift caused by the refraction. If the dispersion is also a problem (in case of broadband high-resolution observations), atmospheric refraction correctors (made from pairs of rotating glass prisms) can be employed as well.

Since the amount of atmospheric refraction is a function of the temperature gradient, temperature, pressure, and humidity (the amount of water vapor, which is especially important at mid-infrared wavelengths), the amount of effort needed for a successful compensation can be prohibitive. Surveyors, on the other hand, will often schedule their observations in the afternoon, when the magnitude of refraction is minimum.

Atmospheric refraction becomes more severe when temperature gradients are strong, and refraction is not uniform when the atmosphere is heterogeneous, as when turbulence occurs in the air. This causes suboptimal seeing conditions, such as the twinkling of stars and various deformations of the Sun's apparent shape soon before sunset or after sunrise.

Astronomical Refraction

Astronomical refraction deals with the angular position of celestial bodies, their appearance as a point source, and through differential refraction, the shape of extended bodies such as the Sun and Moon.

Values

Atmospheric refraction of the light from a star is zero in the zenith, less than 1′ (one arc-minute) at 45° apparent altitude, and still only 5.3′ at 10° altitude; it quickly increases as altitude decreases, reaching 9.9′ at 5° altitude, 18.4′ at 2° altitude, and 35.4′ at the horizon; all values are for 10 °C and 1013.25 hPa in the visible part of the spectrum.

On the horizon refraction is slightly greater than the apparent diameter of the Sun, so when the bottom of the sun's disc appears to touch the horizon, the sun's true altitude is negative. If the atmosphere suddenly vanished at this moment, one couldn't see the sun, as it would be entirely below the horizon. By convention, sunrise and sunset refer to times at which the Sun›s upper limb appears on or disappears from the horizon and the standard value for the Sun›s true altitude is −50′: −34′ for the refraction and −16′ for the Sun's semi-diameter. The altitude of a celestial body is normally given for the center of the body's disc. In the case of the Moon, additional corrections are needed for the Moon's horizontal parallax and its apparent semi-diameter; both vary with the Earth–Moon distance.

Refraction near the horizon is highly variable, principally because of the variability of the temperature gradient near the Earth›s surface and the geometric sensitivity of the nearly horizontal rays to this variability. As early as 1830, Friedrich Bessel had found that even after applying all corrections for temperature and pressure (but not for the temperature gradient) at the observer, highly precise measurements of refraction varied by ±0.19′ at two degrees above the horizon and by ±0.50′ at a half degree above the horizon. At and below the horizon, values of refraction significantly higher than the nominal value of 35.4′ have been observed in a wide range of climates. Georg Constantin Bouris measured refraction of as much of 4° for stars on the horizon at the Athens Observatory and, during his ill-fated Endurance expedition, Sir Ernest Shackleton recorded refraction of 2°37′:

> "The sun which had made 'positively his last appearance' seven days earlier surprised us by lifting more than half its disk above the horizon on May 8. A glow on the northern horizon resolved itself into the sun at 11 am that day. A quarter of an hour later the unreasonable visitor disappeared again, only to rise again at 11:40 am, set at 1 pm, rise at 1:10 pm and set lingeringly at 1:20 pm. These curious phenomena were due to refraction which amounted to 2° 37′ at 1:20 pm. The temperature was 15° below 0° Fahr., and we calculated that the refraction was 2° above normal."

Day-to-day variations in the weather will affect the exact times of sunrise and sunset as well as moon-rise and moon-set, and for that reason it generally is not meaningful to give rise and set times to greater precision than the nearest minute. More precise calculations can be useful for determining day-to-day changes in rise and set times that would occur with the standard value for refraction if it is understood that actual changes may differ because of unpredictable variations in refraction.

Because atmospheric refraction is nominally 34′ on the horizon, but only 29′ at 0.5° above it, the setting or rising sun seems to be flattened by about 5′ (about 1/6 of its apparent diameter).

Calculating Refraction

Young distinguished several regions where different methods for calculating astronomical refraction were applicable. In the upper portion of the sky, with a zenith distance of less than 70° (or an altitude over 20°), various simple refraction formulas based on the index of refraction (and hence on the temperature, pressure, and humidity) at the observer are adequate. Between 20° and 5° of the horizon the temperature gradient becomes the dominant factor and numerical integration, using a method such as that of Auer and Standish and employing the temperature gradient of the standard atmosphere and the measured conditions at the observer, is required. Closer to the horizon, actual measurements of the changes with height of the local temperature gradient need to be employed in the numerical integration. Below the astronomical horizon, refraction is so variable that only crude estimates of astronomical refraction can be made; for example, the observed time of sunrise or sunset can vary by several minutes from day to day. As *The Nautical Almanac* notes, "the actual values of the refraction at low altitudes may, in extreme atmospheric conditions, differ considerably from the mean values used in the tables."

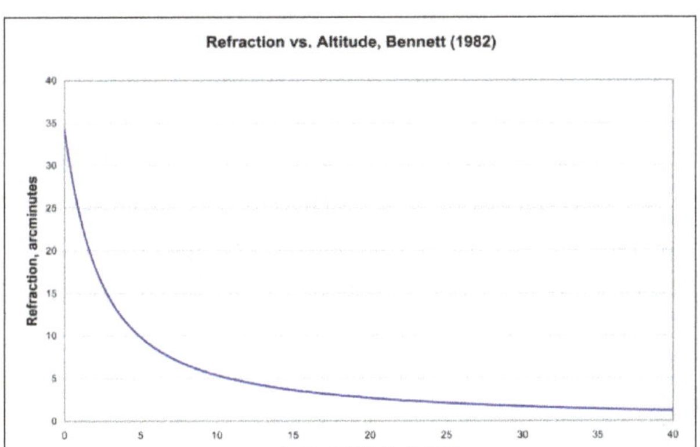

Plot of refraction vs. altitude using Bennett's 1982 formula.

Many different formulas have been developed for calculating astronomical refraction; they are reasonably consistent, differing among themselves by a few minutes of arc at the horizon and becoming increasingly consistent as they approach the zenith. The simpler formulations involved nothing more than the temperature and pressure at the observer, powers of the cotangent of the *apparent* altitude of the astronomical body and in the higher order terms, the height of a fictional homogeneous atmosphere. The simplest version of this formula, which Smart held to be only accurate within 45° of the zenith, is:

$$R = (n_0 - 1)\cot h_a$$

where R is the refraction in radians, n_0 is the index of refraction at the observer (which depends on the temperature and pressure), and h_a is the *apparent* altitude of the astronomical body.

An early simple approximation of this form, which directly incorporated the temperature and pressure at the observer, was developed by George Comstock:

$$R = \frac{21.5b}{273+t}\cot h_a$$

where R is the refraction in seconds of arc, b is the barometric pressure in millimeters of mercury, and t is the Celsiustemperature. Comstock considered that this formula gave results within one arcsecond of Bessel's values for refraction from 15° above the horizon to the zenith.

A further expansion in terms of the third power of the cotangent of the apparent altitude incorporates H_0, the height of the homogeneous atmosphere, in addition to the usual conditions at the observer:

$$R = (n_0 - 1)(1 - H_0)\cot h_a - (n_0 - 1)[H_0 - \frac{1}{2}(n_0 - 1)]\cot^3 h_a$$

A version of this formula is used in the International Astronomical Union's *Standards of Fundamental Astronomy*; a comparison of the IAU's algorithm with more rigorous ray-tracing procedures indicated an agreement within 60 milliarcseconds at altitudes above 15°.

Bennett developed another simple empirical formula for calculating refraction from the apparent altitude which gives the refraction R in arcminutes:

$$R = \cot\left(h_a + \frac{7.31}{h_a + 4.4}\right).$$

This formula is used in the U. S. Naval Observatory's *Vector Astrometry Software*, and is reported to be consistent with Garfinkel's more complex algorithm within 0.07′ over the entire range from the zenith to the horizon. Sæmundssondeveloped an inverse formula for determining refraction from *true* altitude; if h is the true altitude in degrees, refraction R in arcminutes is given by,

$$R = 1.02\cot\left(h + \frac{10.3}{h + 5.11}\right)$$

the formula is consistent with Bennett's to within 0.1′. The formulas of Bennet and Sæmundsson assume an atmospheric pressure of 101.0 kPa and a temperature of 10 °C; for different pressure *P* and temperature *T*, refraction calculated from these formulas is multiplied by,

$$\frac{P}{101}\,\frac{283}{273+T}$$

Refraction increases approximately 1% for every 0.9 kPa increase in pressure, and decreases approximately 1% for every 0.9 kPa decrease in pressure. Similarly, refraction increases approximately 1% for every 3 °C decrease in temperature, and decreases approximately 1% for every 3 °C increase in temperature.

Random Refraction Effects

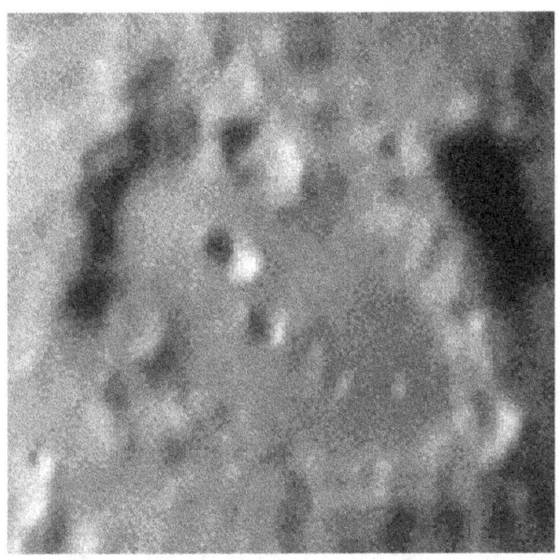

The image of the Moon's surface shows the effects of atmospheric turbulence on the view.

Turbulence in Earth›s atmosphere scatters the light from stars, making them appear brighter and fainter on a time-scale of milliseconds. The slowest components of these fluctuations are visible as twinkling (also called *scintillation*).

Turbulence also causes small, sporadic motions of the star image, and produces rapid distortions in its structure. These effects are not visible to the naked eye, but can be easily seen even in small telescopes. They perturb astronomical seeing conditions. Some telescopes employ adaptive optics to reduce this effect.

Terrestrial Refraction

Terrestrial refraction, sometimes called geodetic refraction, deals with the apparent angular position and measured distance of terrestrial bodies. It is of special concern for the production of precise maps and surveys. Since the line of sight in terrestrial refraction passes near the earth›s surface, the magnitude of refraction depends chiefly on the temperature gradient near the ground, which varies widely at different times of day, seasons of the year, the nature of the terrain, the state of the weather, and other factors.

As a common approximation, terrestrial refraction is considered as a constant bending of the ray of light or line of sight, in which the ray can be considered as describing a circular path. A common measure of refraction is the coefficient of refraction. Unfortunately there are two different definitions of this coefficient. One is the ratio of the radius of the Earth to the radius of the line of sight, the other is the ratio of the angle that the line of sight subtends at the center of the Earth to the angle of refraction measured at the observer. Since the latter definition only measures the bending of the ray at one end of the line of sight, it is one half the value of the former definition.

The coefficient of refraction is directly related to the local vertical temperature gradient and the atmospheric temperature and pressure. The larger version of the coefficient k, measuring the ratio of the radius of the Earth to the radius of the line of sight, is given by:

$$k = 503\frac{P}{T^2}\left(0.0343 + \frac{dT}{dh}\right),$$

where temperature T is given in kelvins, pressure P in millibars, and height h in meters. The angle of refraction increases with the coefficient of refraction and with the length of the line of sight.

Although the straight line from your eye to a distant mountain might be blocked by a closer hill, the ray may curve enough to make the distant peak visible. A convenient method to analyze the effect of refraction on visibility is to consider an increased effective radius of the Earth R_{eff}, given by,

$$R_{eff} = \frac{R}{1 - k},$$

where R is the radius of the Earth and k is the coefficient of refraction. Under this model the ray can be considered a straight line on an Earth of increased radius.

The curvature of the refracted ray in arc seconds per meter can be computed using the relationship,

$$\frac{1}{\sigma} = 16.3\frac{P}{T^2}\left(0.0342 + \frac{dT}{dh}\right)\cos\beta$$

where $1/\sigma$ is the curvature of the ray in arcsec per meter, P is the pressure in millibars, T is the temperature in kelvins, and β is the angle of the ray to the horizontal. Multiplying half the curvature by the length of the ray path gives the angle of refraction at the observer. For a line of sight near the horizon $\cos\beta$ differs little from unity and can be ignored. This yields,

$$= 8.15\frac{LP}{T}\left(0.0342 + \frac{dT}{dh}\right)$$

where L is the length of the line of sight in meters and Ω is the refraction at the observer measured in arc seconds.

A simple approximation is to consider that a mountain's apparent altitude at your eye (in degrees) will exceed its true altitude by its distance in kilometers divided by 1500. This assumes a fairly horizontal line of sight and ordinary air density; if the mountain is very high (so much of the sightline is in thinner air) divide by 1600 instead.

ATMOSPHERIC CIRCULATION

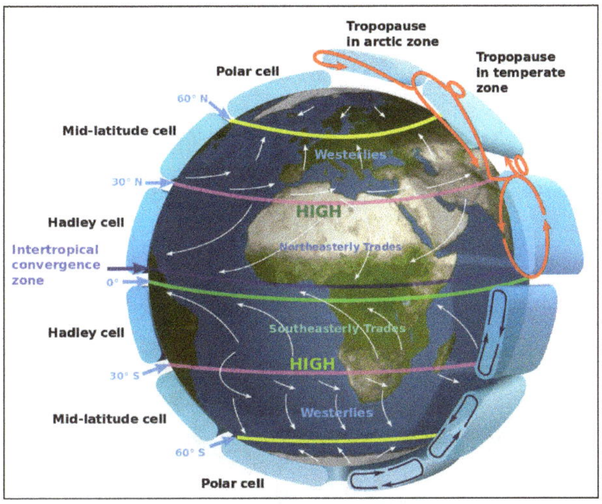

Idealised depiction (at equinox) of large scale atmospheric circulation on Earth.

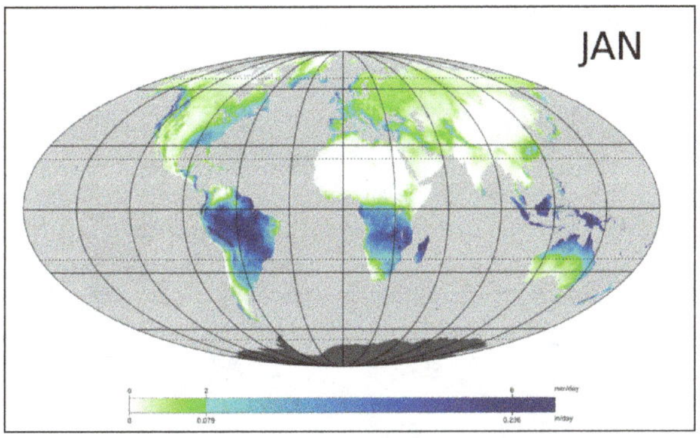

Long-term mean precipitation by month.

Atmospheric circulation is the large-scale movement of air, and together with ocean circulation is the means by which thermal energy is redistributed on the surface of the Earth.

The Earth's atmospheric circulation varies from year to year, but the large-scale structure of its circulation remains fairly constant. The smaller scale weather systems – mid-latitude depressions, or tropical convective cells – occur "randomly", and long-range weather predictions of those cannot be made beyond ten days in practice, or a month in theory.

The Earth's weather is a consequence of its illumination by the Sun, and the laws of thermodynamics. The atmospheric circulation can be viewed as a heat engine driven by the Sun's energy, and whose energy sink, ultimately, is the blackness of space. The work produced by that engine causes the motion of the masses of air and in that process, it redistributes the energy absorbed by the Earth's surface near the tropics to the latitudes nearer the poles, and then to space.

The large-scale atmospheric circulation "cells" shift polewards in warmer periods (for example, interglacials compared to glacials), but remain largely constant as they are, fundamentally, a property of the Earth's size, rotation rate, heating and atmospheric depth, all of which change little. Over very long time periods (hundreds of millions of years), a tectonic uplift can significantly alter their major elements, such as the jet stream, and plate tectonics may shift ocean currents. During the extremely hot climates of the Mesozoic, a third desert belt may have existed at the Equator.

Latitudinal Circulation Features

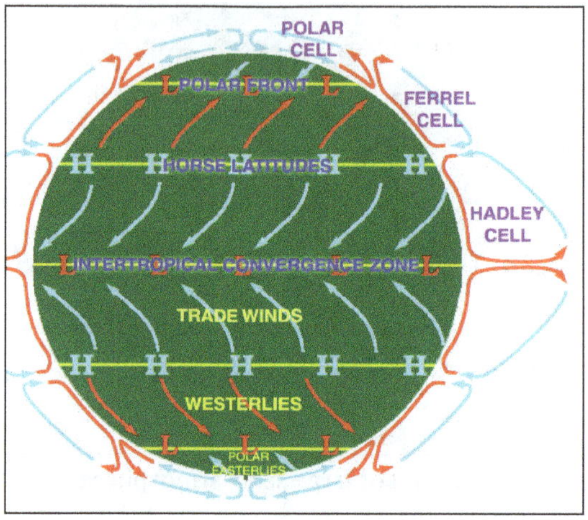

An idealised view of three large circulation cells showing surface winds.

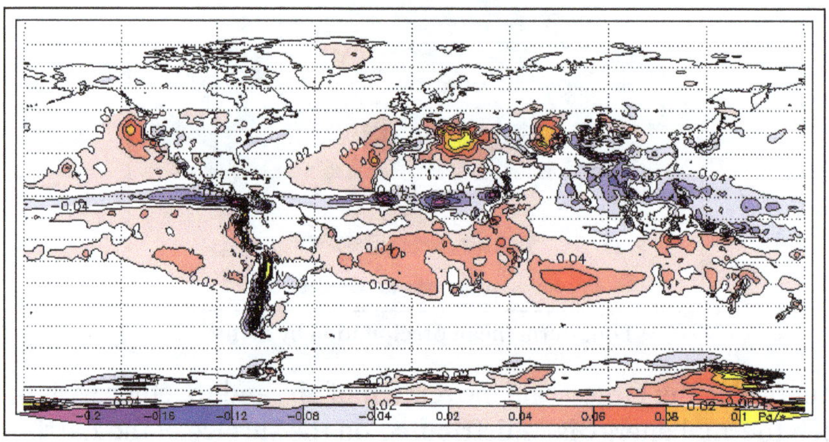

Vertical velocity at 500 hPa, July average. Ascent (negative values) is concentrated close to the solar equator; descent (positive values) is more diffuse but also occurs mainly in the Hadley cell.

The wind belts girdling the planet are organised into three cells in each hemisphere—the Hadley cell, the Ferrel cell, and the polar cell. Those cells exist in both the northern and southern hemispheres. The vast bulk of the atmospheric motion occurs in the Hadley cell. The high pressure systems acting on the Earth's surface are balanced by the low pressure systems elsewhere. As a result, there is a balance of forces acting on the Earth's surface.

The horse latitudes are an area of high pressure at about 30° to 35° latitude (north or south) where winds diverge into the adjacent zones of Hadley or Ferrel cells, and which typically have light winds, sunny skies, and little precipitation.

Hadley Cell

The atmospheric circulation pattern that George Hadley described was an attempt to explain the trade winds. The Hadley cell is a closed circulation loop which begins at the equator. There, moist air is warmed by the Earth's surface, decreases in density and rises. A similar air mass rising on the other side of the equator forces those rising air masses to move poleward. The rising air creates a low pressure zone near the equator. As the air moves poleward, it cools, becomes denser, and descends at about the 30th parallel, creating a high-pressure area. The descended air then travels toward the equator along the surface, replacing the air that rose from the equatorial zone, closing the loop of the Hadley cell. The poleward movement of the air in the upper part of the troposphere deviates toward the east, caused by the coriolis acceleration (a manifestation of conservation of angular momentum). At the ground level, however, the movement of the air toward the equator in the lower troposphere deviates toward the west, producing a wind from the east. The winds that flow to the west (from the east, easterly wind) at the ground level in the Hadley cell are called the Trade Winds.

The ITCZ's band of clouds over the Eastern Pacific and the Americas as seen from space.

Though the Hadley cell is described as located at the equator, in the northern hemisphere it shifts to higher latitudes in June and July and toward lower latitudes in December and January, which is the result of the Sun's heating of the surface. The zone where the greatest heating takes place is called the "thermal equator". As the southern hemisphere summer is December to March, the movement of the thermal equator to higher southern latitudes takes place then.

The Hadley system provides an example of a thermally direct circulation. The power of the Hadley system, considered as a heat engine, is estimated at 200 terawatts.

Ferrel Cell

Part of the air rising at 60° latitude diverges at high altitude toward the poles and creates the polar cell. The rest moves toward the equator where it collides at 30° latitude with the high-level air of the Hadley cell. There it subsides and strengthens the high pressure ridges beneath. A large part of the energy that drives the Ferrel cell is provided by the polar and Hadley cells circulating on either side and that drag the Ferrel cell with it. The Ferrel cell, theorized by William Ferrel (1817–1891), is, therefore, a secondary circulation feature, whose existence depends upon the Hadley and polar cells on either side of it. It might be thought of as an eddy created by the Hadley and polar cells.

The air of the Ferrel cell that descends at 30° latitude returns poleward at the ground level, and as it does so it deviates toward the east. In the upper atmosphere of the Ferrel cell, the air moving toward the equator deviates toward the west. Both of those deviations, as in the case of the Hadley and polar cells, are driven by conservation of angular momentum. As a result, just as the easterly Trade Winds are found below the Hadley cell, the Westerlies are found beneath the Ferrel cell.

The Ferrel cell is weak, because It has neither a strong source of heat nor a strong sink, so the airflow and temperatures within it are variable. For this reason, the mid-latitudes are sometimes known as the "zone of mixing." The Hadley and polar cells are truly closed loops, the Ferrel cell is not, and the telling point is in the Westerlies, which are more formally known as "the Prevailing Westerlies." The easterly Trade Winds and the polar easterlies have nothing over which to prevail, as their parent circulation cells are strong enough and face few obstacles either in the form of massive terrain features or high pressure zones. The weaker Westerlies of the Ferrel cell, however, can be disrupted. The local passage of a cold front may change that in a matter of minutes, and frequently does. As a result, at the surface, winds can vary abruptly in direction. But the winds above the surface, where they are less disrupted by terrain, are essentially westerly. A low pressure zone at 60° latitude that moves toward the equator, or a high pressure zone at 30° latitude that moves poleward, will accelerate the Westerlies of the Ferrel cell. A strong high, moving polewards may bring westerly winds for days.

The Ferrel system acts as a heat pump with a coefficient of performance of 12.1, consuming kinetic energy from the Hadley and polar systems at an approximate rate of 275 terawatts.

Polar Cell

The polar cell is a simple system with strong convection drivers. Though cool and dry relative to equatorial air, the air masses at the 60th parallel are still sufficiently warm and moist to undergo convection and drive a thermal loop. At the 60th parallel, the air rises to the tropopause (about 8 km at this latitude) and moves poleward. As it does so, the upper level air mass deviates toward the east. When the air reaches the polar areas, it has cooled and is considerably denser than the underlying air. It descends, creating a cold, dry high-pressure area. At the polar surface level, the mass of air is driven toward the 60th parallel, replacing the air that rose there, and the polar circulation cell is complete. As the air at the surface moves toward the equator, it deviates toward the west. Again, the deviations of the air masses are the result of the Coriolis effect. The air flows at the surface are called the polar easterlies.

The outflow of air mass from the cell creates harmonic waves in the atmosphere known as Rossby waves. These ultra-long waves determine the path of the polar jet stream, which travels within the transitional zone between the tropopause and the Ferrel cell. By acting as a heat sink, the polar cell moves the abundant heat from the equator toward the polar regions.

The Hadley cell and the polar cell are similar in that they are thermally direct; in other words, they exist as a direct consequence of surface temperatures. Their thermal characteristics drive the weather in their domain. The sheer volume of energy that the Hadley cell transports, and the depth of the heat sink contained within the polar cell, ensures that transient weather phenomena not only have negligible effect on the systems as a whole, but — except under unusual circumstances — they do not form. The endless chain of passing highs and lows which is part of everyday life for mid-latitude dwellers, at latitudes between 30 and 60° latitude, is unknown above the 60th and below the 30th parallels. There are some notable exceptions to this rule. In Europe, unstable weather extends to at least the 70th parallel north.

The polar cell, terrain, and Katabatic winds in Antarctica can create very cold conditions at the surface, for instance the lowest temperature recorded on Earth: −89.2 °C at Vostok Station in Antarctica, measured 1983.

Longitudinal Circulation Features

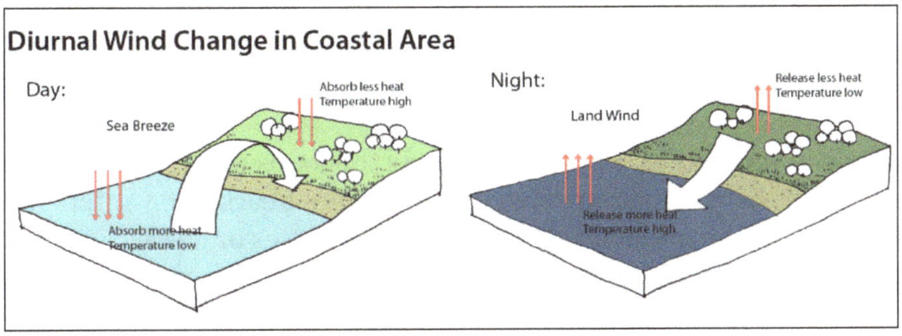

Diurnal wind change in local coastal area, also applies on the continental scale.

While the Hadley, Ferrel, and polar cells (whose axes are oriented along parallels or latitudes) are the major features of global heat transport, they do not act alone. Temperature differences also drive a set of circulation cells, whose axes of circulation are longitudinally oriented. This atmospheric motion is known as zonal overturning circulation.

Latitudinal circulation is a result of the highest solar radiation per unit area (solar intensity) falling on the tropics. The solar intensity decreases as the latitude increases, reaching essentially zero at the poles. Longitudinal circulation, however, is a result of the heat capacity of water, its absorptivity, and its mixing. Water absorbs more heat than does the land, but its temperature does not rise as greatly as does the land. As a result, temperature variations on land are greater than on water.

The Hadley, Ferrel, and polar cells operate at the largest scale of thousands of kilometers (synoptic scale). The latitudinal circulation can also act on this scale of oceans and continents, and this effect is seasonal or even decadal. Warm air rises over the equatorial, continental, and western Pacific Ocean regions. When it reaches the tropopause, it cools and subsides in a region of relatively cooler water mass.

The Pacific Ocean cell plays a particularly important role in Earth's weather. This entirely ocean-based cell comes about as the result of a marked difference in the surface temperatures of the western and eastern Pacific. Under ordinary circumstances, the western Pacific waters are warm, and the eastern waters are cool. The process begins when strong convective activity over equatorial East Asia and subsiding cool air off South America's west coast creates a wind pattern which pushes Pacific water westward and piles it up in the western Pacific. (Water levels in the western Pacific are about 60 cm higher than in the eastern Pacific).

The daily (diurnal) longitudinal effects are at the mesoscale (a horizontal range of 5 to several hundred kilometres). During the day, air warmed by the relatively hotter land rises, and as it does so it draws a cool breeze from the sea that replaces the risen air. At night, the relatively warmer water and cooler land reverses the process, and a breeze from the land, of air cooled by the land, is carried offshore by night.

Walker Circulation

The Pacific cell is of such importance that it has been named the Walker circulation after Sir Gilbert Walker, an early-20th-century director of British observatories in India, who sought a means of predicting when the monsoon winds of India would fail. While he was never successful in doing so, his work led him to the discovery of a link between the periodic pressure variations in the Indian Ocean, and those between the eastern and western Pacific, which he termed the "Southern Oscillation".

The movement of air in the Walker circulation affects the loops on either side. Under normal circumstances, the weather behaves as expected. But every few years, the winters become unusually warm or unusually cold, or the frequency of hurricanes increases or decreases, and the pattern sets in for an indeterminate period.

The Walker Cell plays a key role in this and in the El Niño phenomenon. If convective activity slows in the Western Pacific for some reason (this reason is not currently known), the climates of areas adjacent to the Western Pacific are affected. First, the upper-level westerly winds fail. This cuts off the source of returning, cool air that would normally subside at about 30° south latitude, and therefore the air returning as surface easterlies ceases. There are two consequences. Warm water ceases to surge into the eastern Pacific from the west (it was "piled" by past easterly winds) since there is no longer a surface wind to push it into the area of the west Pacific. This and the corresponding effects of the Southern Oscillation result in long-term unseasonable temperatures and precipitation patterns in North and South America, Australia, and Southeast Africa, and the disruption of ocean currents.

Meanwhile, in the Atlantic, fast-blowing upper level Westerlies of the Hadley cell form, which would ordinarily be blocked by the Walker circulation and unable to reach such intensities. These winds disrupt the tops of nascent hurricanes and greatly diminish the number which are able to reach full strength.

El Niño – Southern Oscillation

El Niño and *La Niña* are opposite surface temperature anomalies of the Southern Pacific, which heavily influence the weather on a large scale. In the case of El Niño, warm surface water approaches the coasts of South America which results in blocking the upwelling of nutrient-rich deep water. This has serious impacts on the fish populations.

In the La Niña case, the convective cell over the western Pacific strengthens inordinately, resulting in colder than normal winters in North America and a more robust cyclone season in South-East Asia and Eastern Australia. There is also an increased upwelling of deep cold ocean waters and more intense uprising of surface air near South America, resulting in increasing numbers of drought occurrences, although fishermen reap benefits from the more nutrient-filled eastern Pacific waters.

ATMOSPHERIC DENSITY

The density of air or atmospheric density, denoted ρ , is the mass per unit volume of Earth's atmosphere. Air density, like air pressure, decreases with increasing altitude. It also changes with variation in atmospheric pressure, temperature and humidity. At 1013.25 hPa (abs) and 15°C, air has a density of approximately 1.225 kg/m³ (0.001225 g/cm³, 0.0023769 slug/(cu ft), 0.0765 lb/(cu ft)) according to ISA (International Standard Atmosphere).

Air density is a property used in many branches of science, engineering, and industry, including aeronautics; gravimetric analysis; the air-conditioning industry; atmospheric research and meteorology; agricultural engineering (modeling and tracking of Soil-Vegetation-Atmosphere-Transfer (SVAT) models); and the engineering community that deals with compressed air.

Depending on the measuring instruments used, different sets of equations for the calculation of the density of air can be applied. Air is a mixture of gases and the calculations always simplify, to a greater or lesser extent, the properties of the mixture.

Dry Air

The density of dry air can be calculated using the ideal gas law, expressed as a function of temperature and pressure:

$$\rho = \frac{p}{R_{specific} T}$$

where,

ρ = air density (kg/m³)

p = absolute pressure (Pa)

T = absolute temperature (K)

$R_{specific}$ = specific gas constant for dry air (J/(kg·K))

The specific gas constant for dry air is 287.058 J/(kg·K) in SI units, and 53.35 (ft·lbf)/(lb·°R) in United States customary and Imperial units. This quantity may vary slightly depending on the molecular composition of air at a particular location.

Therefore:

- At IUPAC standard temperature and pressure (0 °C and 100 kPa), dry air has a density of 1.2754 kg/m³.
- At 20 °C and 101.325 kPa, dry air has a density of 1.2041 kg/m³.
- At 70 °F and 14.696 psi, dry air has a density of 0.074887 lb/ft³.

The following table illustrates the air density–temperature relationship at 1 atm or 101.325 kPa:

Effect of temperature on properties of air			
Temperature T (°C)	Speed of sound c (m/s)	Density of air ρ (kg/m³)	Characteristic specific acoustic impedance z_0 (Pa·s/m)
35	351.88	1.1455	403.2

30	349.02	1.1644	406.5
25	346.13	1.1839	409.4
20	343.21	1.2041	413.3
15	340.27	1.2250	416.9
10	337.31	1.2466	420.5
5	334.32	1.2690	424.3
0	331.30	1.2922	428.0
−5	328.25	1.3163	432.1
−10	325.18	1.3413	436.1
−15	322.07	1.3673	440.3
−20	318.94	1.3943	444.6
−25	315.77	1.4224	449.1

Humid Air

The addition of water vapor to air (making the air humid) reduces the density of the air, which may at first appear counter-intuitive. This occurs because the molar mass of water (18 g/mol) is less than the molar mass of dry air (around 29 g/mol). For any ideal gas, at a given temperature and pressure, the number of molecules is constant for a particular volume. So when water molecules (water vapor) are added to a given volume of air, the dry air molecules must decrease by the same number, to keep the pressure or temperature from increasing. Hence the mass per unit volume of the gas (its density) decreases.

The density of humid air may be calculated by treating it as a mixture of ideal gases. In this case, the partial pressure of water vapor is known as the vapor pressure. Using this method, error in the density calculation is less than 0.2% in the range of −10 °C to 50 °C. The density of humid air is found by:

$$\rho_{\text{humid air}} = \frac{p_d}{R_d T} + \frac{p_v}{R_v T} = \frac{p_d M_d + p_v M_v}{RT}$$

where,

$\rho_{\text{humid air}}$ = Density of the humid air (kg/m³).

p_d = Partial pressure of dry air (Pa).

R_d = Specific gas constant for dry air, 287.058 J/(kg·K).

T = Temperature (K).

p_v = Pressure of water vapor (Pa).

R_v = Specific gas constant for water vapor, 461.495 J/(kg·K).

M_d = Molar mass of dry air, 0.028964 kg/mol.

M_v = Molar mass of water vapor, 0.018016 kg/mol.

R = Universal gas constant, 8.314 J/(K·mol).

The vapor pressure of water may be calculated from the saturation vapor pressure and relative humidity. It is found by:

$$p_v = \phi p_{sat}$$

where,

p_v = Vapor pressure of water

ϕ = Relative humidity

p_{sat} = Saturation vapor pressure

The saturation vapor pressure of water at any given temperature is the vapor pressure when relative humidity is 100%. One formula used to find the saturation vapor pressure is:

$$p_{sat} = 6.102 \times 10^{\frac{7.5T}{T+237.8}}$$

where, T = is in degrees C.

- This equation will give the result of pressure in hPa (100 Pa, equivalent to the older unit millibar; 1 mbar = 0.001 bar = 0.1 kPa)

The partial pressure of dry air p_d is found considering opartial pressure, resulting in:

$$p_d = p - p_v$$

where, p simply denotes the observed absolute pressure.

Variation with Altitude

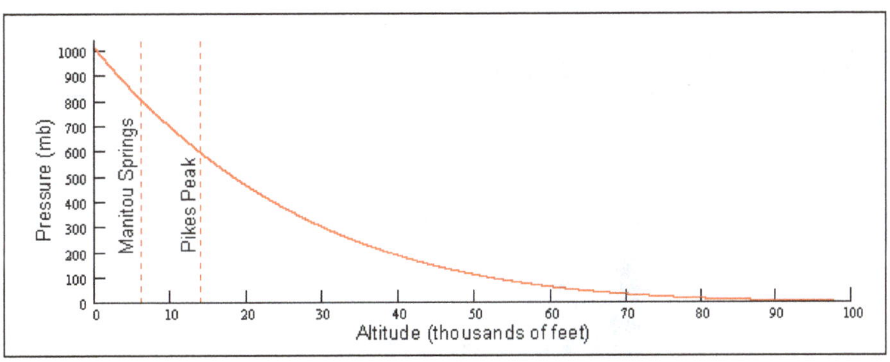

To calculate the density of air as a function of altitude, one requires additional parameters. They are listed below, along with their values according to the International Standard Atmosphere, using for calculation the universal gas constant instead of the air specific constant:

p_0 = sea level standard atmospheric pressure, 101325 Pa

$T_0 =$ sea level standard temperature, 288.15 K

$g =$ earth-surface gravitational acceleration, 9.80665 m/s^2

$L =$ temperature lapse rate, 0.0065 K/m

$R =$ ideal (universal) gas constant, 8.31447 J/(mol·K)

molar mass of dry air, 0.0289644 kg/mol

Temperature at altitude h meters above sea level is approximated by the following formula (only valid inside the troposphere, no more than ~18 km above Earth's surface (and lower away from Equator)):

$$T = T_0 - Lh$$

The pressure at altitude h is given by:

$$p = p_0\left(1 - \frac{Lh}{T_0}\right)^{gM/RL}$$

Density can then be calculated according to a molar form of the ideal gas law:

$$\rho = \frac{pM}{RT}$$

where,

$M =$ molar mass

$R =$ ideal gas constant

$T =$ absolute temperature

$p =$ absolute pressure

DIFFUSE SKY RADIATION

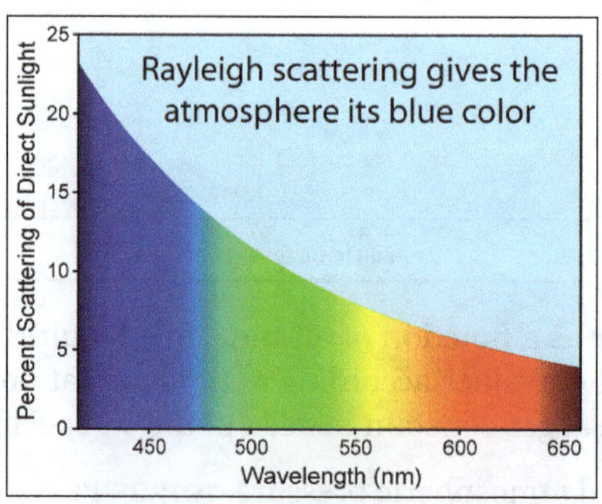

In Earth's atmosphere, the dominant scattering efficiency of blue light is compared to red or green wavelengths. Scattering and absorption are major causes of the attenuation of sunlight radiation by the atmosphere. During broad daylight, the sky is blue due to Rayleigh scattering, while near sunset, and especially during twilight, absorption of irradiation by ozonehelps maintain blue color in the evening sky. At sunrise or sunset, tangentially incident solar rays illuminate clouds with orange-to-red hues.

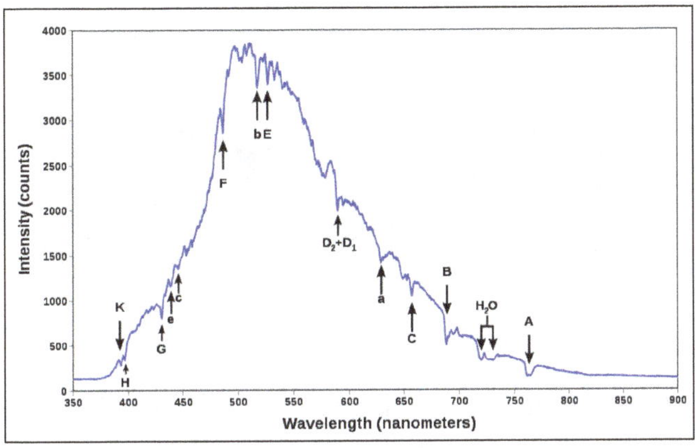

The visible spectrum, about 380 to about 740 nanometers (nm),shows the atmospheric water absorption band and the solar Fraunhofer lines. The blue sky spectrum presents across 450 – 485 nm, the wavelengths of the color blue.

Diffuse sky radiation is solar radiation reaching the Earth's surface after having been scattered from the direct solar beam by molecules or particulates in the atmosphere. Also called sky radiation, diffuse skylight, or just skylight, it is the determinative process for changing the colors of the sky. Approximately 23% of direct incident radiation of total sunlight is removed from the direct solar beam by scattering into the atmosphere; of this amount (of incident radiation) about two-thirds ultimately reaches the earth as photon diffused skylight radiation.

The dominant radiative scattering processes in the atmosphere are Rayleigh scattering and Mie scattering; they are elastic, meaning that a photon of light can be deviated from its path without being absorbed and without changing wavelength.

Under an overcast sky, there is no direct sunlight and all light results from diffused skylight radiation.

Proceeding from analyses of the aftermath of the eruption of the Philippines volcano Mount Pinatubo (in June 1991) and other studies: Diffused skylight, owing to its intrinsic structure and behavior, can illuminate under-canopy leaves, permitting more efficient total whole-plant photosynthesis than would otherwise be the case; this in stark contrast to the effect of totally clear skies with direct sunlight that casts shadows onto understory leaves and thereby limits plant photosynthesis to the top canopy layer.

Color

The Earth's atmosphere scatters short-wavelength light more efficiently than that of longer wavelengths. Because its wavelengths are shorter, blue light is more strongly scattered than the longer-wavelength lights, red or green. Hence the result that when looking at the sky away from the direct incident sunlight, the human eye perceives the sky to be blue. The color perceived is similar to that presented by a monochromatic blue (at wavelength 474–476 nm) mixed with white light, that is, an unsaturated blue light. The explanation of blue color by Rayleigh in 1871 is a famous example of applying dimensional analysis to solving problems in physics.

Clear blue sky.

Scattering and absorption are major causes of the attenuation of sunlight radiation by the atmosphere. Scattering varies as a function of the ratio of particle diameters (of particulates in the atmosphere) to the wavelength of the incident radiation. When this ratio is less than about one-tenth, Rayleigh scattering occurs. (In this case, the scattering coefficient varies inversely with the fourth power of the wavelength. At larger ratios scattering varies in a more complex fashion, as described for spherical particles by the Mie theory.) The laws of geometric optics begin to apply at higher ratios.

Daily at any global venue experiencing sunrise or sunset, most of the solar beam of visible sunlight arrives nearly tangentially to Earth's surface. Here the trajectory of sunlight through the atmosphere is elongated such that much of the blue or green light is scattered away from the line of perceivable visible light. This phenomenon leaves the Sun's rays, and the clouds they illuminate, abundantly orange-to-red in colors, which one sees when looking at a sunset or sunrise.

For the example of the Sun at zenith, during broad daylight, the sky is blue due to Rayleigh scattering, which also involves the diatomic gases N_2 and O_2. Near sunset and especially during twilight, absorption by ozone (O_3) significantly contributes to maintaining blue color in the evening sky.

Neutral Points

There are four commonly detectable points of zero polarization of diffuse sky radiation (known as neutral points) lying along the vertical circle through the sun.

- The Arago point, named after its discoverer, is customarily located at about 20° above the antisolar point; but it lies at higher altitudes in turbid air. The latter property makes the Arago distance a useful measure of atmospheric turbidity.
- The Babinet point, discovered by Jacques Babinet in 1840, is located about 15° to 20° above the sun, hence it is difficult to observe because of solar glare.
- The Brewster point, discovered by David Brewster in 1840, is located about 15° to 20° below the sun; hence it is difficult to observe because of solar glare.

- The fourth point, located at about 20° below the antisolar point, visible only at higher altitudes in the air or in space.

Under an Overcast Sky

There is essentially no direct sunlight under an overcast sky, so all light is then diffuse sky radiation. The flux of light is not very wavelength-dependent because the cloud droplets are larger than the light's wavelength and scatter all colors approximately equally. The light passes through the translucent clouds in a manner similar to frosted glass. The intensity ranges (roughly) from $1/6$ of direct sunlight for relatively thin clouds down to $1/1000$ of direct sunlight under the extreme of thickest storm clouds.

As a Part of Total Radiation

One of the equations for total solar radiation is:

$$H_t = H_b R_b + H_d R_d + (H_b + H_d) R_r$$

where H_b is the beam radiation irradiance, R_b is the tilt factor for beam radiation, H_d is the diffuse radiation irradiance, R_d is the tilt factor for diffuse radiation and R_r is the tilt factor for reflected radiation.

R_b is given by:

$$R_b = \frac{\sin(\delta)\sin(\phi - \beta) + \cos(\delta)\cos(h)\cos(\phi - \beta)}{\sin(\delta)\sin(\phi) + \cos(\delta)\cos(h)\cos(\phi)}$$

where δ is the solar declination, Φ is the latitude, β is an angle from the horizontal and h is the solar hour angle.

R_d is given by:

$$R_d = \frac{1 + \cos(\beta)}{2}$$

and R_r by:

$$R_r = \frac{\rho(1 - \cos(\beta))}{2}$$

where ρ is the reflectivity of the surface.

Agriculture and the Eruption of Mt. Pinatubo

The eruption of the Philippines volcano - Mount Pinatubo in June 1991 ejected roughly 10 km^3 (2.4 cu mi) of magma and "17,000,000 metric tons"(17 teragrams) of sulfur dioxide SO_2 into the air, introducing ten times as much total SO_2 as the 1991 Kuwaiti fires, mostly during the explosive

Plinian/Ultra-Plinian event of June 15, 1991, creating a global stratospheric SO_2 haze layer which persisted for years. This resulted in the global average temperature dropping by about 0.5 °C (0.9 °F). As volcanic ash falls out of the atmosphere rapidly, the negative agricultural effects of the eruption were largely immediate and localized to a relatively small area in close proximity to the eruption, as they were caused by the resulting thick ash cover that resulted. Globally however, despite a several-month 5% drop in overall solar irradiation, and a reduction in direct sunlight by 30%, there was no negative impact to global agriculture. Surprisingly, a 3-4 yearincrease in global Agricultural productivity and forestry growth was observed, excepting boreal forest regions.

A Space Shuttle (Mission STS-43) photograph of the Earth over South America taken on August 8, 1991, which captures the double layer of Pinatubo aerosol clouds (dark streaks) above lower cloud tops.

Under more-or-less direct sunlight, dark shadows that limit photosynthesis are castonto understorey leaves. Within the thicket, very little direct sunlight can enter.

The means by which this was discovered, is that initially at the time, a mysterious drop in the rate at which carbon dioxide (CO_2) was filling the atmosphere was observed, which is charted in what is known as the "Keeling Curve". This led numerous scientists to assume that this reduction was due to the lowering of the Earth›s temperature, and with that, a slow down in plant and soil respiration, indicating a deleterious impact to global agriculture from the volcanic haze layer. However upon actual investigation, the reduction in the rate at which carbon dioxide filled the atmosphere did not match up with the hypothesis that plant respiration rates had declined. Instead the advantageous anomaly was relatively firmly linked to an unprecedented increase in the growth/ net primary production, of global plant life, resulting in the increase of the carbon sink effect of global photosynthesis. The mechanism by which the increase in plant growth was possible, was that the 30% reduction of direct sunlight can also be expressed as an increase or "enhancement" in the amount of diffuse sunlight.

The Diffused Skylight Effect

Well lit understorey areas due to overcast clouds creating diffuse/soft sunlight conditions, that permits photosynthesis on leaves under the canopy.

This diffused skylight, owing to its intrinsic nature, can illuminate under-canopy leaves permitting more efficient total whole-plant photosynthesis than would otherwise be the case. In stark contrast to the effect of totally clear skies and the direct sunlight that results from it, which casts shadows onto understorey leaves, strictly limiting plant photosynthesis to the top canopy layer. This increase in global agriculture from the volcanic haze layer also naturally results as a product of other aerosols that are not emitted by volcanoes, such as man-made "moderately thick smoke loading" pollution, as the same mechanism, the "aerosol direct radiative effect" is behind both.

ATMOSPHERIC ELECTRICITY

Cloud to ground lightning. Typically, lightning discharges 30,000 amperes, at up to 100 million volts, and emits light, radio waves, x-rays and even gamma rays. Plasma temperatures in lightning can approach 28,000 kelvins.

Atmospheric electricity is the study of electrical charges in the Earth›s atmosphere (or that of another planet). The movement of charge between the Earth's surface, the atmosphere, and the ionosphere is known as the global atmospheric electrical circuit. Atmospheric electricity is an interdisciplinary topic with a long history, involving concepts from electrostatics, atmospheric physics, meteorology and Earth science.

Thunderstorms act as a giant battery in the atmosphere, charging up the ionosphere to about 400,000 volts with respect to the surface. This sets up an electric field throughout the atmosphere, which decreases with increase in altitude. Atmospheric ions created by cosmic rays and natural radioactivity move in the electric field, so a very small current flows through the atmosphere, even away from thunderstorms. Near the surface of the earth, the magnitude of the field is on average around 100 V/m.

Atmospheric electricity involves both thunderstorms, which create lightning bolts to rapidly discharge huge amounts of atmospheric charge stored in storm clouds, and the continual electrification of the air due to ionization from cosmic rays and natural radioactivity, which ensure that the atmosphere is never quite neutral.

Atmospheric electricity is always present, and during fine weather away from thunderstorms, the air above the surface of Earth is positively charged, while the Earth's surface charge is negative. It can be understood in terms of a difference of potentialbetween a point of the Earth's surface, and a point somewhere in the air above it. Because the atmospheric electric field is negatively directed in fair weather, the convention is to refer to the potential gradient, which has the opposite sign and is about 100V/m at the surface. The potential gradient in most locations is much lower than this value because it is an average of the charge built up by every thunderstorm and atmospheric disturbance around the globe. There is a weak conduction current of atmospheric ions moving in the atmospheric electric field, about 2 picoAmperes per square metre, and the air is weakly conductive due to the presence of these atmospheric ions.

Variations

Global daily cycles in the atmospheric electric field, with a minimum around 03 UT and peaking roughly 16 hours later, were researched by the Carnegie Institution of Washington in the 20th century. This Carnegie curve variation has been described as "the fundamental electrical heartbeat of the planet".

Even away from thunderstorms, atmospheric electricity can be highly variable, but, generally, the electric field is enhanced in fogs and dust whereas the atmospheric electrical conductivity is diminished.

Near Space

The electrosphere layer (from tens of kilometers above the surface of the earth to the ionosphere) has a high electrical conductivity and is essentially at a constant electric potential. The ionosphere is the inner edge of the magnetosphere and is the part of the atmosphere that is ionized by solar radiation. (Photoionization is a physical process in which a photon is incident on an atom, ion or molecule, resulting in the ejection of one or more electrons.)

Cosmic Radiation

The Earth, and almost all living things on it, are constantly bombarded by radiation from outer space. This radiation primarily consists of positively charged ions from protons to iron and larger nuclei derived sources outside our solar system. This radiation interacts with atoms in the atmosphere to create an air shower of secondary ionising radiation, including X-rays, muons, protons, alpha particles, pions, and electrons. Ionization from this secondary radiation ensures that the atmosphere is weakly conductive, and that the slight current flow from these ions over the Earth's surface balances the current flow from thunderstorms. Ions have characteristic parameters such as mobility, lifetime, and generation rate that vary with altitude.

Thunderstorms and Lightning

The potential difference between the ionosphere and the Earth is maintained by thunderstorms, with lightning strikes delivering negative charges from the atmosphere to the ground.

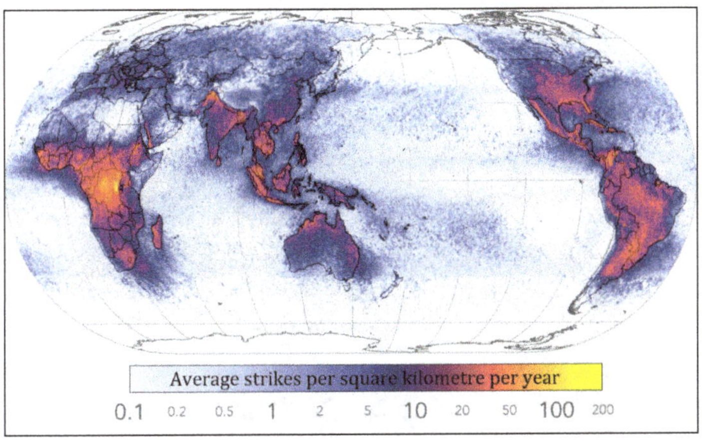

World map showing frequency of lightning strikes, in flashes per km² per year (equal-area projection). Lightning strikes most frequently in the Democratic Republic of the Congo. Combined 1995–2003 data from the Optical Transient Detector and 1998–2003 data from the Lightning Imaging Sensor.

Collisions between ice and soft hail (graupel) inside cumulonimbus clouds causes separation of positive and negative charges within the cloud, essential for the generation of lightning. How lightning initially forms is still a matter of debate: Scientists have studied root causes ranging from atmospheric perturbations (wind, humidity, and atmospheric pressure) to the impact of solar wind and energetic particles.

An average bolt of lightning carries a negative electric current of 40 kiloamperes (kA)(although some bolts can be up to 120 kA), and transfers a charge of five coulombsand energy of 500 MJ, or enough energy to power a 100-watt lightbulb for just under two months. The voltage depends on the length of the bolt, with the dielectric breakdown of air being three million volts per meter, and lightning bolts often being several hundred meters long. However, lightning leader development is not a simple matter of dielectric breakdown, and the ambient electric fields required for lightning leader propagation can be a few orders of magnitude less than dielectric breakdown strength. Further, the potential gradient inside a well-developed return-stroke channel is on the order of hundreds of volts per meter or less due to intense channel ionization, resulting in a true power output on the order of megawatts per meter for a vigorous return-stroke current of 100 kA.

If the quantity of water that is condensed in and subsequently precipitated from a cloud is known, then the total energy of a thunderstorm can be calculated. In an average thunderstorm, the energy released amounts to about 10,000,000 kilowatt-hours (3.6×10^{13} joule), which is equivalent to a 20-kiloton nuclear warhead. A large, severe thunderstorm might be 10 to 100 times more energetic.

Lightning sequence (Duration: 0.32 seconds).

Corona Discharges

A depiction of atmospheric electricity in a Martian dust storm, which has been suggested as a possible explanation for enigmatic chemistry results from Mars.

St. Elmo's Fire is an electrical phenomenon in which luminous plasma is created by a coronal dischargeoriginating from a grounded object. Ball lightning is often erroneously identified as St. Elmo›s Fire, whereas they are separate and distinct phenomena. Although referred to as "fire", St. Elmo›s Fire is, in fact, plasma, and is observed, usually during a thunderstorm, at the tops of trees, spires or other tall objects, or on the heads of animals, as a brush or star of light.

Corona is caused by the electric field around the object in question ionizing the air molecules, producing a faint gloweasily visible in low-light conditions. Approximately 1,000 – 30,000 volts per centimetre is required to induce St. Elmo's Fire; however, this is dependent on the geometryof the object in question. Sharp points tend to require lower voltage levels to produce the same result because electric fields are more concentrated in areas of high curvature, thus discharges are more intense at the end of pointed objects. St. Elmo's Fire and normal sparks both can appear when high electrical voltage affects a gas. St. Elmo's fire is seen during thunderstorms when the ground below the storm is electrically charged, and there is high voltage in the air between the cloud and the ground. The voltage tears apart the air molecules and the gas begins to glow. The nitrogen and oxygen in the Earth's atmosphere causes St. Elmo's Fire to fluoresce with blue or violet light; this is similar to the mechanism that causes neon signs to glow.

Earth-ionosphere Cavity

The Schumann resonances are a set of spectrum peaks in the extremely low frequency (ELF) portion of the Earth's electromagnetic field spectrum. Schumann resonance is due to the space between the surface of the Earth and the conductive ionosphere acting as a waveguide. The limited dimensions of the earth cause this waveguide to act as a resonant cavity for electromagnetic waves. The cavity is naturally excited by energy from lightning strikes.

Electrical System Grounding

Atmospheric charges can cause undesirable, dangerous, and potentially lethal charge potential buildup in suspended electric wire power distribution systems. Bare wires suspended in the air spanning many kilometers and isolated from the ground can collect very large stored charges at high voltage, even when there is no thunderstorm or lightning occurring. This charge will seek to discharge itself through the path of least insulation, which can occur when a person reaches out to activate a power switch or to use an electric device.

To dissipate atmospheric charge buildup, one side of the electrical distribution system is connected to the earth at many points throughout the distribution system, as often as on every support pole. The one earth-connected wire is commonly referred to as the "protective earth", and provides path for the charge potential to dissipate without causing damage, and provides redundancy in case any one of the ground paths is poor due to corrosion or poor ground conductivity. The additional electric grounding wire that carries no power serves a secondary role, providing a high-current short-circuit path to rapidly blow fuses and render a damaged device safe, rather than have an ungrounded device with damaged insulation become "electrically live" via the grid power supply, and hazardous to touch.

Each transformer in an alternating current distribution grid segments the grounding system into a new separate circuit loop. These separate grids must also be grounded on one side to prevent charge buildup within them relative to the rest of the system, and which could cause damage from charge potentials discharging across the transformer coils to the other grounded side of the distribution network.

References

- Starr, cecie (2005). Biology: concepts and applications. Thomson brooks/cole. Isbn 978-0-534-46226-0
- Atmospheric-temperature, encyclopedias-almanacs-transcripts-and-maps, science: encyclopedia.com, Retrieved 21 April, 2019
- "Atmospheric pressure (encyclopedic entry)". National geographic. Retrieved 28 february 2018.
- "Q & a: pressure - gravity matters?". Department of physics. University of illinois urbana-champaign. Retrieved 28 february 2018
- World: highest sea level air pressure below 750 m, wmo.asu.edu, 1968-12-31, archived from the original on 2013-05-14, retrieved 2013-04-15
- Us department of commerce, national oceanic and atmospheric administration. "what are the horse latitudes?". Oceanservice.noaa.gov. Retrieved 2019-04-14
- Yochanan kushnir (2000). "the climate system: general circulation and climate zones". Retrieved 13 march 2012

6

Weather Forecasting

The application of science and technology to predict the atmospheric conditions of a given location and time is referred to as weather forecasting. Nowcasting, marine weather forecasting, numerical weather prediction, etc. are some of the various types of weather forecasting. This chapter discusses in detail these types of weather forecasting.

Weather forecasting is the prediction of the weather through application of the principles of physics, supplemented by a variety of statistical and empirical techniques. In addition to predictions of atmospheric phenomena themselves, weather forecasting includes predictions of changes on Earth's surface caused by atmospheric conditions—e.g., snow and ice cover, storm tides, and floods.

Measurements and Ideas as the basis for Weather Prediction

The observations of few other scientific enterprises are as vital or affect as many people as those related to weather forecasting. From the days when early humans ventured from caves and other natural shelters, perceptive individuals in all likelihood became leaders by being able to detect nature's signs of impending snow, rain, or wind, indeed of any change in weather. With such information they must have enjoyed greater success in the search for food and safety, the major objectives of that time.

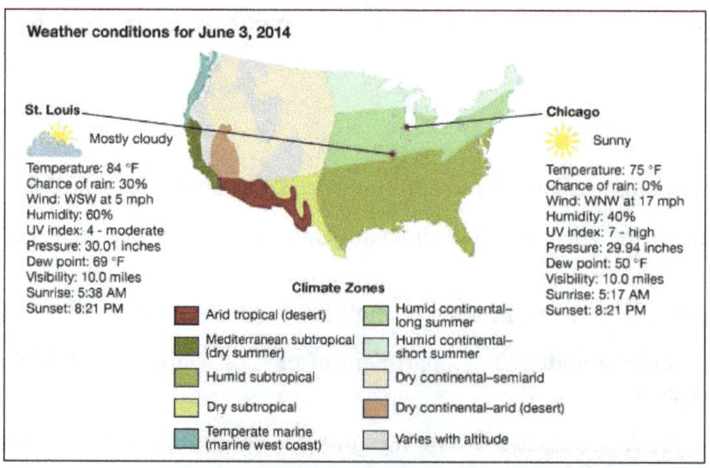

Weather conditions Comparing weather conditions for St. Louis and Chicago on a given day.

In a sense, weather forecasting is still carried out in basically the same way as it was by the earliest humans—namely, by making observations and predicting changes. The modern tools used to measure temperature, pressure, wind, and humidity in the 21st century would certainly amaze them, and the results obviously are better. Yet, even the most sophisticated numerically calculated forecast made on a supercomputer requires a set of measurements of the condition of the atmosphere—an initial picture of temperature, wind, and other basic elements, somewhat comparable to that formed by our forebears when they looked out of their cave dwellings. The primeval approach entailed insights based on the accumulated experience of the perceptive observer, while the modern technique consists of solving equations. Although seemingly quite different, there are underlying similarities between both practices. In each case the forecaster asks "What is?" in the sense of "What kind of weather prevails today?" and then seeks to determine how it will change in order to extrapolate what it will be.

Because observations are so critical to weather prediction, an account of meteorological measurements and weather forecasting is a story in which ideas and technology are closely intertwined, with creative thinkers drawing new insights from available observations and pointing to the need for new or better measurements, and technology providing the means for making new observations and for processing the data derived from measurements. The basis for weather prediction started with the theories of the ancient Greek philosophers and continued with Renaissance scientists, the scientific revolution of the 17th and 18th centuries, and the theoretical models of 20th- and 21st-century atmospheric scientists and meteorologists. Likewise, it tells of the development of the "synoptic" idea—that of characterizing the weather over a large region at exactly the same time in order to organize information about prevailing conditions. In synoptic meteorology, simultaneous observations for a specific time are plotted on a map for a broad area whereby a general view of the weather in that region is gained. The so-called synoptic weather map came to be the principal tool of 19th-century meteorologists and continues to be used today in weather stations and on television weather reports around the world.

Since the mid-20th century, digital computers have made it possible to calculate changes in atmospheric conditions mathematically and objectively—i.e., in such a way that anyone can obtain the same result from the same initial conditions. The widespread adoption of numerical weather prediction models brought a whole new group of players—computer specialists and experts in numerical processing and statistics—to the scene to work with atmospheric scientists and meteorologists. Moreover, the enhanced capability to process and analyze weather data stimulated the long-standing interest of meteorologists in securing more observations of greater accuracy. Technological advances since the 1960s led to a growing reliance on remote sensing, particularly the gathering of data with specially instrumented Earth-orbiting satellites. By the late 1980s, forecasts of the weather were largely based on the determinations of numerical models integrated by high-speed supercomputers—except for some shorter-range predictions, particularly those related to local thunderstorm activity, which were made by specialists directly interpreting radar and satellite measurements. By the early 1990s a network of next-generation Doppler weather radar (NEXRAD) was largely in place in the United States, which allowed meteorologists to predict severe weather events with additional lead time before their occurrence. During the late 1990s and early 21st century, computer processing power increased, which allowed weather bureaus to produce more-sophisticated ensemble forecasts—that is, sets of multiple model runs whose results limit the range of uncertainty with respect to a forecast.

Practical Applications

Systematic weather records were kept after instruments for measuring atmospheric conditions became available during the 17th century. Undoubtedly these early records were employed mainly by those engaged in agriculture. Planting and harvesting obviously can be planned better and carried out more efficiently if long-term weather patterns can be estimated. In the United States, the foundations of the national weather services were laid down by American physicist Joseph Henry, the first head of the Smithsonian Institution. In 1849 Henry created a network of volunteer weather observers to help improve storm prediction in the U.S. The first national weather services were provided by the U.S. Army Signal Corps beginning on February 9, 1870, which also incorporated Henry's volunteer weather observers by 1874. These operations were taken over by the Department of Agriculture in 1891. By the early 1900s free mail service and telephone were providing forecasts daily to millions of American farmers. The U.S. Weather Bureau established a Fruit-Frost (forecasting) Service during World War I, and by the 1920s radio broadcasts to agricultural interests were being made in most states.

Weather forecasting became an important tool for aviation during the 1920s and '30s. Its application in this area gained in importance after Francis W. Reichelderfer was appointed chief of the U.S. Weather Bureau (USWB) in 1939. Reichelderfer had previously modernized the U.S. Navy's meteorological service and made it a model of support for naval aviation. During World War II the discovery of very strong wind currents at high altitudes (the jet streams, which can affect aircraft speed) and the general susceptibility of military operations in Europe to weather led to a special interest in weather forecasting.

One of the most famous wartime forecasting problems was for Operation Overlord, the invasion of the European mainland at Normandy by Allied forces. An unusually intense June storm brought high seas and gales to the French coast, but a moderation of the weather that was successfully predicted by Col. J.M. Stagg of the British forces (after consultation with both British and American forecasters) enabled Gen. Dwight D. Eisenhower, supreme commander of the Allied Expeditionary Forces, to make his critical decision to invade on June 6, 1944.

The second half of the 20th century saw a reorganization of the country's weather bureau. The USWB was part of the Department of Agriculture until 1940, when it was added to the Department of Commerce. On October 9, 1970, the USWB became the National Weather Service.

In addition, the later part of the 20th century was a time of unprecedented growth of commercial weather-forecasting firms in the United States and elsewhere. Marketing organizations and stores hire weather-forecasting consultants to help with the timing of sales and promotions of products ranging from snow tires and roofing materials to summer clothes and resort vacations. Many oceangoing shipping vessels as well as military ships use optimum ship routing forecasts to plan their routes in order to minimize lost time, potential damage, and fuel consumption in heavy seas. Similarly, airlines carefully consider atmospheric conditions when planning long-distance flights so as to avoid the strongest head winds and to ride with the strongest tail winds.

International trading of foodstuffs such as wheat, corn (maize), beans, sugar, cocoa, and coffee can be severely affected by weather news. For example, in 1975 a severe freeze in Brazil caused the price of coffee to increase substantially within just a few weeks, and in 2017 Georgia peach growers blamed the combination of warm winter temperatures and a spring freeze on the loss of nearly 80 percent of the state's peach crop. In addition, extreme heat and drought can affect production; one study estimated that 9–10 percent of cereal crops between 1964 and 2007 were lost to these phenomena. Weather-forecasting organizations are thus frequently called upon by banks, commodity traders, and food companies to give them advance knowledge of the possibility of such sudden changes. The cost of all sorts of commodities and services, whether they are tents for outdoor events or plastic covers for the daily newspapers, can be reduced or eliminated if reliable information about possible precipitation can be obtained in advance.

Forecasts must be quite precise for applications that are tailored to specific industries. Gas and electric utilities, for example, may require forecasts of temperature within one or two degrees a day ahead of time, or ski-resort operators may need predictions of nighttime relative humidity on the slopes within 5 to 10 percent in order to schedule snow making.

Early Measurements and Ideas

The Greek philosophers had much to say about meteorology, and many who subsequently engaged in weather forecasting no doubt made use of their ideas. Unfortunately, they probably made many bad forecasts, because Aristotle, who was the most influential, did not believe that wind is air in motion. He did believe, however, that west winds are cold because they blow from the sunset.

The scientific study of meteorology did not develop until measuring instruments became available. Its beginning is commonly associated with the invention of the mercury barometer by Evangelista Torricelli, an Italian physicist-mathematician, in the mid-17th century and the nearly concurrent development of a reliable thermometer. (Galileo had constructed an elementary form of gas thermometer in 1607, but it was defective; the efforts of many others finally resulted in a reasonably accurate liquid-in-glass device.)

Torricelli, EvangelistaItalian physicist and mathematician Evangelista Torricelli, inventor of the mercury barometer.

A succession of notable achievements by chemists and physicists of the 17th and 18th centuries contributed significantly to meteorological research. The formulation of the laws of gas pressure, temperature, and density by Robert Boyle and Jacques-Alexandre-César Charles, the development of calculus by Isaac Newton and Gottfried Wilhelm Leibniz, the development of the law of partial pressures of mixed gases by John Dalton, and the formulation of the doctrine of latent heat (i.e., heat release by condensation or freezing) by Joseph Black are just a few of the major scientific breakthroughs of the period that made it possible to measure and better understand theretofore unknown aspects of the atmosphere and its behaviour. During the 19th century, all of these brilliant ideas began to produce results in terms of useful weather forecasts.

The Emergence of Synoptic Forecasting Methods

Analysis of Synoptic Weather Reports

An observant person who has learned nature's signs can interpret the appearance of the sky, the wind, and other local effects and "foretell the weather." A scientist can use instruments at one location to do so even more effectively. The modern approach to weather forecasting, however, can only be realized when many such observations are exchanged quickly by experts at various weather stations and entered on a synoptic weather map to depict the patterns of pressure, wind, temperature, clouds, and precipitation at a specific time. Such a rapid exchange of weather data became feasible with the development of the electric telegraph in 1837 by Samuel F.B. Morse of the United States. By 1849 Joseph Henry of the Smithsonian Institution in Washington, D.C., was plotting daily weather maps based on telegraphic reports, and in 1869 Cleveland Abbe at the Cincinnati Observatory began to provide regular weather forecasts using data received telegraphically.

Synoptic weather maps resolved one of the great controversies of meteorology—namely, the rotary storm dispute. By the early decades of the 19th century, it was known that storms were associated with low barometric readings, but the relation of the winds to low-pressure systems, called cyclones, remained unrecognized. William Redfield, a self-taught meteorologist from Middletown, Conn., noticed the pattern of fallen trees after a New England hurricane and suggested in 1831 that the wind flow was a rotary counterclockwise circulation around the centre of lowest pressure. The American meteorologist James P. Espy subsequently proposed in his Philosophy of Storms (1841) that air would flow toward the regions of lowest pressure and then would be forced upward, causing clouds and precipitation. Both Redfield and Espy proved to be right. The air does spin around the cyclone, as Redfield believed, while the layers close to the ground flow inward and upward as well. The net result is a rotational wind circulation that is slightly modified at Earth's surface to produce inflow toward the storm centre, just as Espy had proposed. Further, the inflow is associated with clouds and precipitation in regions of low pressure, though that is not the only cause of clouds there.

In Europe the writings of Heinrich Dove, a Polish scientist who directed the Prussian Meteorological Institute, greatly influenced views concerning wind behaviour in storms. Unlike the Americans, Dove did not focus on the pattern of the winds around the storm but rather on how the wind should change at one place as a storm passed. It was many years before his followers understood the complexity of the possible changes.

Establishment of Weather-Station Networks and Services

Routine production of synoptic weather maps became possible after networks of stations were organized to take measurements and report them to some type of central observatory. As early as 1814, U.S. Army Medical Corps personnel were ordered to record weather data at their posts; this activity was subsequently expanded and made more systematic. Actual weather-station networks were established in the United States by New York University, the Franklin Institute, and the Smithsonian Institution during the early decades of the 19th century.

In Britain, James Glaisher organized a similar network, as did Christophorus H.D. Buys Ballot in the Netherlands. Other such networks of weather stations were developed near Vienna, Paris, and St. Petersburg.

It was not long before national meteorological services were established on the Continent and in the United Kingdom. The first national weather service in the United States commenced operations in 1871, with responsibility assigned to the U.S. Army Signal Corps. The original purpose of the service was to provide storm warnings for the Atlantic and Gulf coasts and for the Great Lakes. Within the next few decades, national meteorological services were established in such countries as Japan, India, and Brazil. The importance of international cooperation in weather prognostication was recognized by the directors of such national services. By 1880 they had formed the International Meteorological Organization (IMO).

The proliferation of weather-station networks linked by telegraphy made synoptic forecasting a reality by the close of the 19th century. Yet, the daily weather forecasts generated left much to be desired. Many errors occurred as predictions were largely based on the experience that each individual forecaster had accumulated over several years of practice, vaguely formulated rules of thumb (e.g., of how pressure systems move from one region to another), and associations that were poorly understood, if at all.

Progress During the Early 20th Century

An important aspect of weather prediction is to calculate the atmospheric pressure pattern—the positions of the highs and lows and their changes. Modern research has shown that sea-level pressure patterns respond to the motions of the upper-atmospheric winds, with their narrow, fast-moving jet streams and waves that propagate through the air and pass air through themselves.

Frequent surprises and errors in estimating surface atmospheric pressure patterns undoubtedly caused 19th-century forecasters to seek information about the upper atmosphere for possible explanations. The British meteorologist Glaisher made a series of ascents by balloon during the 1860s, reaching an unprecedented height of nine kilometres. At about this time investigators on the Continent began using unmanned balloons to carry recording barographs, thermographs, and hygrographs to high altitudes. During the late 1890s meteorologists in both the United States and Europe used kites equipped with instruments to probe the atmosphere up to altitudes of about three kilometres. Notwithstanding these efforts, knowledge about the upper atmosphere remained very limited at the turn of the century. The situation was aggravated by the confusion created by observations from weather stations located on mountains or hilltops. Such observations often did not show what was expected, partly because so little was known about the upper atmosphere and partly because the mountains themselves affect measurements, producing results that are not representative of what would be found in the free atmosphere at the same altitude.

Fortunately, a large enough number of scientists had already put forth ideas that would make it possible for weather forecasters to think three-dimensionally, even if sufficient meteorological measurements were lacking. Henrik Mohn, the first of a long line of highly creative Norwegian meteorologists, Wladimir Köppen, the noted German climatologist, and Max Margules, an influential Russian-born meteorologist, all contributed to the view that mechanisms of the upper air generate the energy of storms.

In 1911 William H. Dines, a British meteorologist, published data that showed how the upper atmosphere compensates for the fact that the low-level winds carry air toward low-pressure centres. Dines recognized that the inflow near the ground is more or less balanced by a circulation upward and outward aloft. Indeed, for a cyclone to intensify, which would require a lowering of central pressure, the outflow must exceed the inflow; the surface winds can converge quite strongly toward the cyclone, but sufficient outflow aloft can produce falling pressure at the centre.

Meteorologists of the time were now aware that vertical circulations and upper-air phenomena were important, but they still had not determined how such knowledge could improve weather forecasting. Then, in 1919, the Norwegian meteorologist Jacob Bjerknes introduced what has been referred to as the Norwegian cyclone model. This theory pulled together many earlier ideas and related the patterns of wind and weather to a low-pressure system that exhibited fronts—which are rather sharp sloping boundaries between cold and warm air masses. Bjerknes pointed out the rainfall/snowfall patterns that are characteristically associated with the fronts in cyclones: the rain or snow occurs over large areas on the cold side of an advancing warm front poleward of a low-pressure centre. Here, the winds are from the lower latitudes, and the warm air, being light, glides up over a large region of cold air. Widespread, sloping clouds spread ahead of the cyclone; barometers fall as the storm approaches, and precipitation from the rising warm air falls through the cold air below. Where the cold air advances to the rear of the storm, squalls and showers mark the abrupt lifting of the warm air being displaced. Thus, the concept of fronts focused attention on the action at air mass boundaries. The Norwegian cyclone model could be called the frontal model, for the idea of warm air masses being lifted over cold air along their edges (fronts) became a major forecasting tool. The model not only emphasized the idea but it also showed how and where to apply it.

In later work, Bjerknes and several other members of the so-called Bergen school of meteorology expanded the model to show that cyclones grow from weak disturbances on fronts, pass through a regular life cycle, and ultimately die by the inflow filling them. Both the Norwegian cyclone model and the associated life-cycle concept are still used today by weather forecasters.

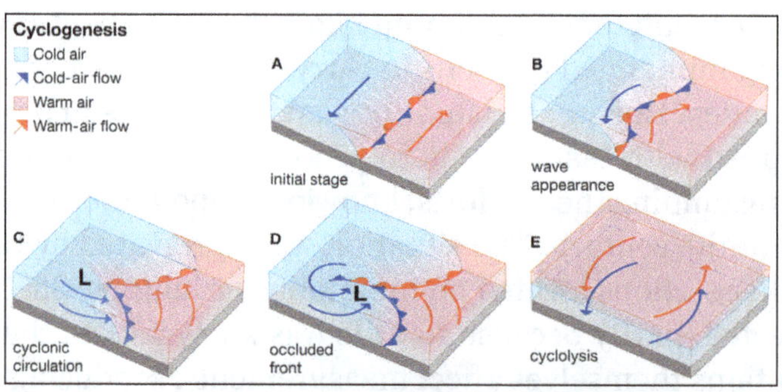

Evolution of a wave (frontal) cyclone.

While Bjerknes and his Bergen colleagues refined the cyclone model, other Scandinavian meteorologists provided much of the theoretical basis for modern weather prediction. Foremost among them were Vilhelm Bjerknes, Jacob's father, and Carl-Gustaf Rossby. Their ideas helped make it possible to understand and carefully calculate the changes in atmospheric circulation and the motion of the upper-air waves that control the behaviour of cyclones.

Modern trends and Developments

Upper-air Observations by Means of Balloon-borne Sounding Equipment

Once again technology provided the means with which to test the new scientific ideas and stimulate yet newer ones. During the late 1920s and '30s, several groups of investigators (those headed by Yrjö Väisälä of Finland and Pavel Aleksandrovich Malchanov of the Soviet Union, for example) began using small radio transmitters with balloon-borne instruments, eliminating the need to recover the instruments and speeding up access to the upper-air data. These radiosondes, as they came to be called, gave rise to the upper-air observation networks that still exist today. Approximately 75 stations in the United States and more than 500 worldwide release, twice daily, balloons that reach heights of 30,000 metres or more. Observations of temperature and relative humidity at various pressures are radioed back to the station from which the balloons are released as they ascend at a predetermined rate. The balloons also are tracked by radar and global positioning system (GPS) satellites to ascertain the behaviour of winds from their drift.

Forecasters are able to produce synoptic weather maps of the upper atmosphere twice each day on the basis of radiosonde observations. While new methods of upper-air measurement have been developed, the primary synoptic clock times for producing upper-air maps are still the radiosonde-observation times—namely, 0000 (midnight) and 1200 (noon) Greenwich Mean Time (GMT). Furthermore, modern computer-based forecasts use 0000 and 1200 GMT as the starting times from which they calculate the changes that are at the heart of modern forecasts. It is, in effect, the synoptic approach carried out in a different way, intimately linked to the radiosonde networks developed during the 1930s and '40s.

Application of Radar

As in many fields of endeavour, weather prediction experienced several breakthroughs during and immediately after World War II. The British began using microwave radar in the late 1930s to monitor enemy aircraft, but it was soon learned that radar gave excellent returns from raindrops at certain wavelengths (5 to 10 centimetres). As a result it became possible to track and study the evolution of individual showers or thunderstorms, as well as to "see" the precipitation structure of larger storms.

Since its initial application in meteorological work, radar has grown as a forecaster's tool. Virtually all tornadoes and severe thunderstorms over the United States and in some other parts of the world are monitored by radar. Radar observation of the growth, motion, and characteristics of such storms provide clues as to their severity. Modern radar systems use the Doppler principle of frequency shift associated with movement toward or away from the radar transmitter/receiver to determine wind speeds as well as storm motions.

Using radar and other observations, the Japanese American meteorologist Tetsuya Theodore Fujita discovered many details of severe thunderstorm behaviour and of the structure of the violent local storms common to the Midwest region of the United States. His Doppler-radar analyses of winds revealed "microburst" gusts. These gusts cause the large wind shears (differences) associated with strong rains that have been responsible for some plane crashes.

The well-defined eye and the rain bands of Hurricane Hyacinth about 805 kilometres (500 miles) south of the southern tip of Baja California, Mexico, photographed from an Earth-orbiting satellite.

Other types of radar have been used increasingly for detecting winds continuously, as opposed to twice a day. These wind-profiling radar systems actually pick up signals "reflected" by clear air and so can function even when no clouds or rain are present.

Meteorological Measurements from Satellites and Aircraft

A major breakthrough in meteorological measurement came with the launching of the first meteorological satellite, the TIROS (Television and Infrared Observation Satellite), by the United States on April 1, 1960. The impact of global quantitative views of temperature, cloud, and moisture distributions, as well as of surface properties (e.g., ice cover and soil moisture), has already been substantial. Furthermore, new ideas and new methods may very well make the 21st century the "age of the satellite" in weather prediction.

Medium-range forecasts that provide information five to seven days in advance were impossible before satellites began making global observations—particularly over the ocean waters of the Southern Hemisphere—routinely available in real time. Global forecasting models developed at the U.S. National Center for Atmospheric Research (NCAR), the European Centre for Medium Range Weather Forecasts (ECMWF), and the U.S. National Meteorological Center (NMC) became the standard during the 1980s, making medium-range forecasting a reality. Global weather forecasting models are routinely run by national weather services around the world, including those of Japan, the United Kingdom, and Canada.

Meteorological satellites travel in various orbits and carry a wide variety of sensors. They are of two principal types: the low-flying polar orbiter, and the geostationary orbiter.

The first type circle Earth at altitudes of 500–1,000 kilometres and in roughly north–south orbits. They appear overhead at any one locality twice a day and provide very high-resolution data because they fly close to Earth. Such satellites are vitally necessary for much of Europe and other high-latitude locations because they orbit near the poles. These satellites do, however, suffer from one major limitation: they can provide a sampling of atmospheric conditions only twice daily.

The geostationary satellite is made to orbit Earth along its equatorial plane at an altitude of about 36,000 kilometres. At that height the eastward motion of the satellite coincides exactly with Earth's rotation, so that the satellite remains in one position above the Equator. Satellites of this type are able to provide an almost continuous view of a wide area. Because of this capability, geostationary satellites have yielded new information about the rapid changes that occur in thunderstorms, hurricanes, and certain types of fronts, making them invaluable to weather forecasting as well as meteorological research.

One weakness common to virtually all satellite-borne sensors and to some ground-based radars that use UHF/VHF waves is an inability to measure thin layers of the atmosphere. One such layer is the tropopause, the boundary between the relatively dry stratosphere and the more meteorologically active layer below. This is often the region of the jet streams. Important information about these kinds of high-speed air currents is obtained with sensors mounted on high-flying commercial aircraft and is routinely included in global weather analyses.

Numerical Weather Prediction (NWP) Models

Thinkers frequently advance ideas long before the technology exists to implement them. Few better examples exist than that of numerical weather forecasting. Instead of mental estimates or rules of thumb about the movement of storms, numerical forecasts are objective calculations of changes to the weather map based on sets of physics-based equations called models. Shortly after World War I a British scientist named Lewis F. Richardson completed such a forecast that he had been working on for years by tedious and difficult hand calculations. Although the forecast proved to be incorrect, Richardson's general approach was accepted decades later when the electronic computer became available. In fact, it has become the basis for nearly all present-day weather forecasts. Human forecasters may interpret or even modify the results of the computer models, but there are few forecasts that do not begin with numerical-model calculations of pressure, temperature, wind, and humidity for some future time.

The method is closely related to the synoptic approach. Data are collected rapidly by a Global Telecommunications System for 0000 or 1200 GMT to specify the initial conditions. The model equations are then solved for various segments of the weather map—often a global map—to calculate how much conditions are expected to change in a given time, say, 10 minutes. With such changes added to the initial conditions, a new map is generated (in the computer's memory) valid for 0010 or 1210 GMT. This map is treated as a new set of initial conditions, probably not quite as accurate as the measurements for 0000 and 1200 GMT but still very accurate. A new step is undertaken to generate a forecast for 0020 or 1220. This process is repeated step after step. In principle, the process could continue indefinitely. In practice, small errors creep into the calculations, and they accumulate. Eventually, the errors become so large by this cumulative process that there is no point in continuing.

Global numerical forecasts are produced regularly (once or twice daily) at the ECMWF, the NMC, and the U.S. military facilities in Omaha, Neb., and Monterey, Calif., and in Tokyo, Moscow, London, Melbourne, and elsewhere. In addition, specialized numerical forecasts designed to predict more details of the weather are made for many smaller regions of the world by various national weather services, military organizations, and even a few private companies. Finally, research versions of numerical weather prediction models are constantly under review, development, and testing at NCAR and at the Goddard Space Flight Center in the United States and at universities in several nations.

The capacity and complexity of numerical weather prediction models have increased dramatically since the mid-1940s when the earliest modeling work was done by the mathematician John von Neumann and the meteorologist Jule Charney at the Institute for Advanced Study in Princeton, N.J. Because of their pioneering work and the discovery of important simplifying relationships by other scientists (notably Arnt Eliassen of Norway and Reginald Sutcliffe of Britain), a joint U.S. Weather Bureau, Navy, and Air Force numerical forecasting unit was formed in 1954 in Washington, D.C. Referred to as JNWP, this unit was charged with producing operational numerical forecasts on a daily basis.

The era of numerical weather prediction thus really began in the 1950s. As computing power grew, so did the complexity, speed, and capacity for detail of the models. And as new observations became available from such sources as Earth-orbiting satellites, radar systems, and drifting weather balloons, so too did methods sophisticated enough to ingest the data into the models as improved initial synoptic maps.

Numerical forecasts have improved steadily over the years. The vast Global Weather Experiment, first conceived by Charney, was carried out by many nations in 1979 under the leadership of the World Meteorological Organization to demonstrate what high-quality global observations could do to improve forecasting by numerical prediction models. The results of that effort continue to effect further improvement.

A relatively recent development has been the construction of mesoscale numerical prediction models. The prefix meso- means "middle" and here refers to middle-sized features in the atmosphere, between large cyclonic storms and individual clouds. Fronts, clusters of thunderstorms, sea breezes, hurricane bands, and jet streams are mesoscale structures, and their evolution and behaviour are crucial forecasting problems that only recently have been dealt with in numerical prediction. An example of such a model is the meso-eta model, which was developed by Serbian atmospheric scientist Fedor Mesinger and Serbian-born American atmospheric scientist Zaviša Janjić. The meso-eta model is a finer-scale version of a regional numerical weather prediction model used by the National Weather Service in the United States. The national weather services of several countries produce numerical forecasts of considerable detail by means of such limited-area mesoscale models.

Principles and Methodology of Weather Forecasting

Short-range Forecasting

When people wait under a shelter for a downpour to end, they are making a very-short-range weather forecast. They are assuming, based on past experience, that such hard rain usually does not last very long. In short-term predictions the challenge for the forecaster is to improve on what the layperson can do. For years the type of situation represented in the above example proved particularly vexing for forecasters, but since the mid-1980s they have been developing a method called nowcasting to meet precisely this sort of challenge. In this method, radar and satellite observations of local atmospheric conditions are processed and displayed rapidly by computers to project weather several hours in advance. The U.S. National Oceanic and Atmospheric Administration operates a facility known as PROFS (Program for Regional Observing and Forecasting Services) in Boulder, Colo., specially equipped for nowcasting.

Meteorologists can make somewhat longer-term forecasts (those for 6, 12, 24, or even 48 hours) with considerable skill because they are able to measure and predict atmospheric conditions for large areas by computer. Using models that apply their accumulated expert knowledge quickly, accurately, and in a statistically valid form, meteorologists are now capable of making forecasts objectively. As a consequence, the same results are produced time after time from the same data inputs, with all analysis accomplished mathematically. Unlike the prognostications of the past made with subjective methods, objective forecasts are consistent and can be studied, reevaluated, and improved.

Another technique for objective short-range forecasting is called MOS (for Model Output Statistics). Conceived by Harry R. Glahn and D.A. Lowry of the U.S. National Weather Service, this method involves the use of data relating to past weather phenomena and developments to extrapolate the values of certain weather elements, usually for a specific location and time period. It overcomes the weaknesses of numerical models by developing statistical relations between model forecasts and observed weather. These relations are then used to translate the model forecasts directly to specific weather forecasts. For example, a numerical model might not predict the occurrence of surface winds at all, and whatever winds it did predict might always be too strong. MOS relations can automatically correct for errors in wind speed and produce quite accurate forecasts of wind occurrence at a specific point, such as Heathrow Airport near London. As long as numerical weather prediction models are imperfect, there may be many uses for the MOS technique.

Predictive Skills and Procedures

Short-range weather forecasts generally tend to lose accuracy as forecasters attempt to look farther ahead in time. Predictive skill is greatest for periods of about 12 hours and is still quite substantial for 48-hour predictions. An increasingly important group of short-range forecasts are economically motivated. Their reliability is determined in the marketplace by the economic gains they produce (or the losses they avert).

Weather warnings are a special kind of short-range forecast; the protection of human life is the forecaster's greatest challenge and source of pride. The first national weather forecasting service in the United States (the predecessor of the Weather Bureau) was in fact formed, in 1870, in response to the need for storm warnings on the Great Lakes. Increase Lapham of Milwaukee urged Congress to take action to reduce the loss of hundreds of lives incurred each year by Great Lakes shipping during the 1860s. The effectiveness of the warnings and other forecasts assured the future of the American public weather service.

Weather warnings are issued by government and military organizations throughout the world for all kinds of threatening weather events: tropical storms variously called hurricanes, typhoons, or tropical cyclones, depending on location; great oceanic gales outside the tropics spanning hundreds of kilometres and at times packing winds comparable to those of tropical storms; and, on land, flash floods, high winds, fog, blizzards, ice, and snowstorms.

A particular effort is made to warn of hail, lightning, and wind gusts associated with severe thunderstorms, sometimes called severe local storms (SELS) or simply severe weather. Forecasts and warnings also are made for tornadoes, those intense, rotating windstorms that represent the most violent end of the weather scale. Destruction of property and the risk of injury and death are extremely high in the path of a tornado, especially in the case of the largest systems (sometimes called maxi-tornadoes).

Because tornadoes are so uniquely life-threatening and because they are so common in various regions of the United States, the National Weather Service operates a National Severe Storms Forecasting Center (NSSFC) in Kansas City, Mo., where SELS forecasters survey the atmosphere for the conditions that can spawn tornadoes or severe thunderstorms. This group of SELS forecasters, assembled in 1952, monitors temperature and water vapour in an effort to identify the warm, moist regions where thunderstorms may form and studies maps of pressure and winds to find regions where the storms may organize into mesoscale structures. The group also monitors jet streams and dry air aloft that can combine to distort ordinary thunderstorms into rare rotating ones with tilted chimneys of upward rushing air that, because of the tilt, are unimpeded by heavy falling rain. These high-speed updrafts can quickly transport vast quantities of moisture to the cold upper regions of the storms, thereby promoting the formation of large hailstones. The hail and rain drag down air from aloft to complete a circuit of violent, cooperating updrafts and downdrafts.

By correctly anticipating such conditions, SELS forecasters are able to provide time for the mobilization of special observing networks and personnel. If the storms actually develop, specific warnings are issued based on direct observations. This two-step process consists of the tornado or severe thunderstorm watch, which is the forecast prepared by the SELS forecaster, and the warning, which is usually released by a local observing facility. The watch may be issued when the skies are clear, and it usually covers a number of counties. It alerts the affected area to the threat but does not attempt to pinpoint which communities will be affected.

By contrast, the warning is very specific to a locality and calls for immediate action. Radar of various types can be used to detect the large hailstones, the heavy load of raindrops, the relatively clear region of rapid updraft, and even the rotation in a tornado. These indicators, or an actual sighting, often trigger the tornado warning. In effect, a warning is a specific statement that danger is imminent, whereas a watch is a forecast that warnings may be necessary later in a given region.

Long-range Forecasting

Techniques

Extended-range, or long-range, weather forecasting has had a different history and a different approach from short- or medium-range forecasting. In most cases, it has not applied the synoptic method of going forward in time from a specific initial map. Instead, long-range forecasters have tended to use the climatological approach, often concerning themselves with the broad weather picture over a period of time rather than attempting to forecast day-to-day details.

There is good reason to believe that the limit of day-to-day forecasts based on the "initial map" approach is about two weeks. Most long-range forecasts thus attempt to predict the departures from normal conditions for a given month or season. Such departures are called anomalies. A forecast might state that "spring temperatures in Minneapolis have a 65 percent probability of being above normal." It would likely be based on a forecast anomaly map, which shows temperature anomaly patterns. The maps do not attempt to predict the weather for a particular day, but rather forecast trends (i.e., warmer than normal) for an extended amount of time, such as a season (i.e., spring).

The U.S. Weather Bureau began making experimental long-range forecasts just before the beginning of World War II, and its successor, the National Weather Service, continues to express such predictions in probabilistic terms, making it clear that they are subject to uncertainty. Verification shows that forecasts of temperature anomalies are more reliable than those of precipitation, that monthly forecasts are better than seasonal ones, and that winter months are predicted somewhat more accurately than other seasons.

Prior to the 1980s the technique commonly used in long-range forecasting relied heavily on the analog method, in which groups of weather situations (maps) from previous years were compared to those of the current year to determine similarities with the atmosphere's present patterns (or "habits"). An association was then made between what had happened subsequently in those "similar" years and what was going to happen in the current year. Most of the techniques were quite subjective, and there were often disagreements of interpretation and consequently uneven quality and marginal reliability.

Persistence (warm summers follow warm springs) or anti-persistence (cold springs follow warm winters) also were used, even though, strictly speaking, most forecasters consider persistence forecasts "no-skill" forecasts. Yet, they too have had limited success.

Prospects for New Procedures

In the last quarter of the 20th century the approach of and prospects for long-range weather forecasting changed significantly. Stimulated by the work of Jerome Namias, who headed the U.S. Weather Bureau's Long-Range Forecast Division for 30 years, scientists began to look at ocean-surface temperature anomalies as a potential cause for the temperature anomalies of the atmosphere in succeeding seasons and at distant locations. At the same time, other American meteorologists, most notably John M. Wallace, showed how certain repetitive patterns of atmospheric flow were related to each other in different parts of the world. With satellite-based observations available, investigators began to study the El Niño phenomenon. Atmospheric scientists also revived the work of Gilbert Walker, an early 20th-century British climatologist who had studied the Southern Oscillation, the aforementioned up-and-down fluctuation of atmospheric pressure in the Southern Hemisphere. Walker had investigated related air circulations (later called the Walker Circulation) that resulted from abnormally high pressures in Australia and low pressures in Argentina or vice versa.

All of this led to new knowledge about how the occurrence of abnormally warm or cold ocean waters and of abnormally high or low atmospheric pressures could be interrelated in vast global connections. Knowledge about these links—El Niño/Southern Oscillation (ENSO)—and about the behaviour of parts of these vast systems enables forecasters to make better long-range predictions, at least in part, because the ENSO features change slowly and somewhat regularly. This approach of studying interconnections between the atmosphere and the ocean may represent the beginning of a revolutionary stage in long-range forecasting.

Since the mid-1980s, interest has grown in applying numerical weather prediction models to long-range forecasting. In this case, the concern is not with the details of weather predicted 20 or 30 days in advance but rather with objectively predicted anomalies. The reliability of long-range forecasts, like that of short- and medium-range projections, has improved substantially in recent years. Yet, many significant problems remain unsolved, posing interesting challenges for all those engaged in the field.

NUMERICAL WEATHER PREDICTION

Weather models use systems of differential equations based on the laws of physics, which are in detail fluid motion, thermodynamics, radiative transfer, and chemistry, and use a coordinate system which divides the planet into a 3D grid. Winds, heat transfer, solar radiation, relative humidity, phase changes of water and surface hydrology are calculated within each grid cell, and the interactions with neighboring cells are used to calculate atmospheric properties in the future.

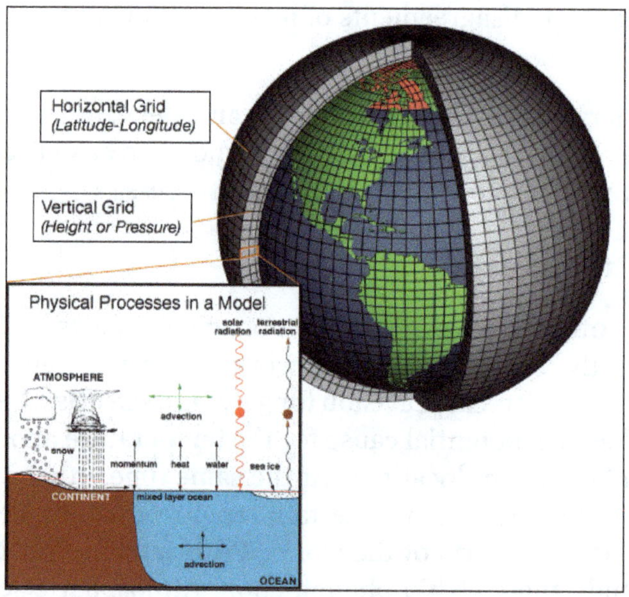

Numerical weather prediction (NWP) uses mathematical models of the atmosphere and oceans to predict the weather based on current weather conditions. Though first attempted in the 1920s, it was not until the advent of computer simulation in the 1950s that numerical weather predictions produced realistic results. A number of global and regional forecast models are run in different countries worldwide, using current weather observations relayed from radiosondes, weather satellites and other observing systems as inputs.

Mathematical models based on the same physical principles can be used to generate either short-term weather forecasts or longer-term climate predictions; the latter are widely applied for understanding and projecting climate change. The improvements made to regional models have allowed for significant improvements in tropical cyclone track and air quality forecasts; however, atmospheric models perform poorly at handling processes that occur in a relatively constricted area, such as wildfires.

Manipulating the vast datasets and performing the complex calculations necessary to modern numerical weather prediction requires some of the most powerful supercomputers in the world. Even with the increasing power of supercomputers, the forecast skill of numerical weather models extends to only about six days. Factors affecting the accuracy of numerical predictions include the density and quality of observations used as input to the forecasts, along with deficiencies in the numerical models themselves. Post-processing techniques such as model output statistics (MOS) have been developed to improve the handling of errors in numerical predictions.

A more fundamental problem lies in the chaotic nature of the partial differential equations that govern the atmosphere. It is impossible to solve these equations exactly, and small errors grow with time (doubling about every five days). Present understanding is that this chaotic behavior limits accurate forecasts to about 14 days even with perfectly accurate input data and a flawless model. In addition, the partial differential equations used in the model need to be supplemented with parameterizations for solar radiation, moist processes (clouds and precipitation), heat exchange, soil, vegetation, surface water, and the effects of terrain. In an effort to quantify the large amount of inherent uncertainty remaining in numerical predictions, ensemble forecasts have been used since the 1990s to help gauge the confidence in the forecast, and to obtain useful results farther into the future than otherwise possible. This approach analyzes multiple forecasts created with an individual forecast model or multiple models.

Initialization

Weather reconnaissance aircraft, such as this WP-3D Orion, provide data that is then used in numerical weather forecasts.

The atmosphere is a fluid. As such, the idea of numerical weather prediction is to sample the state of the fluid at a given time and use the equations of fluid dynamics and thermodynamics to estimate the state of the fluid at some time in the future. The process of entering observation data into the model to generate initial conditions is called *initialization*. On land, terrain maps available at resolutions down to 1 kilometer (0.6 mi) globally are used to help model atmospheric circulations within regions of rugged topography, in order to better depict features such as downslope winds, mountain waves and related cloudiness that affects incoming solar radiation. The main inputs from country-based weather services are observations from devices (called radiosondes) in weather balloons that measure various atmospheric parameters and transmits them to a fixed receiver, as well as from weather satellites. The World Meteorological Organization acts to standardize the instrumentation, observing practices and timing of these observations worldwide. Stations either report hourly in METARreports, or every six hours in SYNOP reports. These observations are irregularly spaced, so they are processed by data assimilation and objective analysis methods, which perform quality control and obtain values at locations usable by the model›s mathematical algorithms. The data are then used in the model as the starting point for a forecast.

A variety of methods are used to gather observational data for use in numerical models. Sites launch radiosondes in weather balloons which rise through the troposphere and well into the stratosphere. Information from weather satellites is used where traditional data sources are not available. Commerce provides pilot reports along aircraft routes and ship reports along shipping routes. Research projects use reconnaissance aircraft to fly in and around weather systems of interest, such as tropical cyclones. Reconnaissance aircraft are also flown over the open oceans during the cold season into systems which cause significant uncertainty in forecast guidance, or are expected to be of high impact from three to seven days into the future over the downstream continent. Sea ice began to be initialized in forecast models in 1971. Efforts to involve sea surface temperature in model initialization began in 1972 due to its role in modulating weather in higher latitudes of the Pacific.

Computation

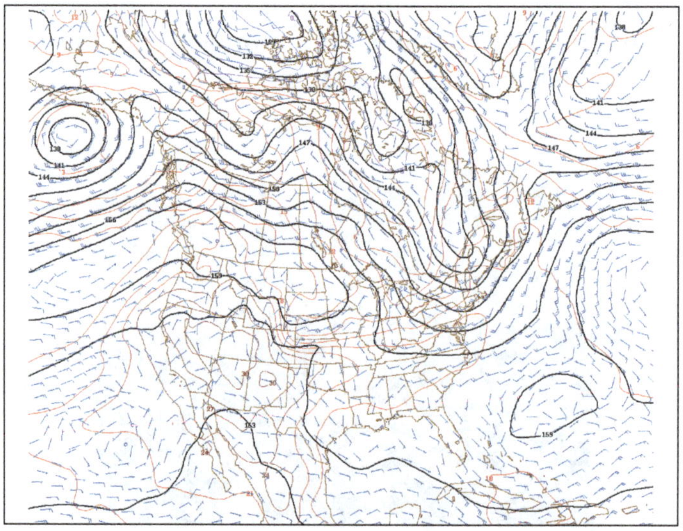

A prognostic chart of the 96-hour forecast of 850 mbar geopotential height and temperaturefrom the Global Forecast System.

An atmospheric model is a computer program that produces meteorologicalinformation for future times at given locations and altitudes. Within any modern model is a set of equations, known as the primitive equations, used to predict the future state of the atmosphere. These equations—along with the ideal gas law—are used to evolve the density, pressure, and potential temperature scalar fields and the air velocity (wind) vector field of the atmosphere through time. Additional transport equations for pollutants and other aerosols are included in some primitive-equation high-resolution models as well. The equations used are nonlinear partial differential equations which are impossible to solve exactly through analytical methods, with the exception of a few idealized cases. Therefore, numerical methods obtain approximate solutions. Different models use different solution methods: some global models and almost all regional models use finite difference methods for all three spatial dimensions, while other global models and a few regional models use spectral methods for the horizontal dimensions and finite-difference methods in the vertical.

These equations are initialized from the analysis data and rates of change are determined. These rates of change predict the state of the atmosphere a short time into the future; the time increment for this prediction is called a *time step*. This future atmospheric state is then used as the starting point for another application of the predictive equations to find new rates of change, and these new rates of change predict the atmosphere at a yet further time step into the future. This time stepping is repeated until the solution reaches the desired forecast time. The length of the time step chosen within the model is related to the distance between the points on the computational grid, and is chosen to maintain numerical stability. Time steps for global models are on the order of tens of minutes, while time steps for regional models are between one and four minutes. The global models are run at varying times into the future. The UKMET Unified Model is run six days into the future, while the European Centre for Medium-Range Weather Forecasts' Integrated Forecast System and Environment Canada's Global Environmental Multiscale Model both run out to ten days into the future, and the Global Forecast Systemmodel run by the Environmental Modeling Center is run sixteen days into the future. The visual output produced by a model solution is known as a prognostic chart, or prog.

Parameterization

Field of cumulus clouds, which are parameterized since they are too small to be explicitly included within numerical weather prediction.

Some meteorological processes are too small-scale or too complex to be explicitly included in numerical weather prediction models. *Parameterization* is a procedure for representing these processes by relating them to variables on the scales that the model resolves. For example, the gridboxes in weather and climate models have sides that are between 5 kilometers (3 mi) and 300 kilometers (200 mi) in length. A typical cumulus cloud has a scale of less than 1 kilometer (0.6 mi), and would require a grid even finer than this to be represented physically by the equations of fluid motion. Therefore, the processes that such clouds represent are parameterized, by processes of various sophistication. In the earliest models, if a column of air within a model gridbox was conditionally unstable (essentially, the bottom was warmer and moister than the top) and the water vapor content at any point within the column became saturated then it would be overturned (the warm, moist air would begin rising), and the air in that vertical column mixed. More sophisticated schemes recognize that only some portions of the box might convect and that entrainment and other processes occur. Weather models that have gridboxes with sizes between 5 and 25 kilometers (3 and 16 mi) can explicitly represent convective clouds, although they need to parameterize cloud microphysics which occur at a smaller scale. The formation of large-scale (stratus-type) clouds is more physically based; they form when the relative humidity reaches some prescribed value.

Sub-grid scale processes need to be taken into account. Rather than assuming that clouds form at 100% relative humidity, the cloud fraction can be related to a critical value of relative humidity less than 100%, reflecting the sub grid scale variation that occurs in the real world.

The amount of solar radiation reaching the ground, as well as the formation of cloud droplets occur on the molecular scale, and so they must be parameterized before they can be included in the model. Atmospheric drag produced by mountains must also be parameterized, as the limitations in the resolution of elevation contours produce significant underestimates of the drag.This method of parameterization is also done for the surface flux of energy between the ocean and the atmosphere, in order to determine realistic sea surface temperatures and type of sea ice found near the ocean's surface. Sun angle as well as the impact of multiple cloud layers is taken into account. Soil type, vegetation type, and soil moisture all determine how much radiation goes into warming and how much moisture is drawn up into the adjacent atmosphere, and thus it is important to parameterize their contribution to these processes. Within air quality models, parameterizations take into account atmospheric emissions from multiple relatively tiny sources (e.g. roads, fields, factories) within specific grid boxes.

Domains

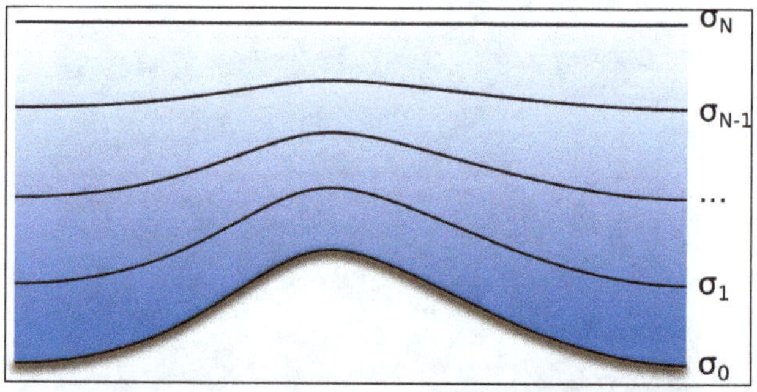

A cross-section of the atmosphere over terrain with a sigma-coordinate representation shown. Mesoscale models divide the atmosphere vertically using representations similar to the one shown here.

The horizontal domain of a model is either *global*, covering the entire Earth, or *regional*, covering only part of the Earth. Regional models (also known as *limited-area* models, or LAMs) allow for the use of finer grid spacing than global models because the available computational resources are focused on a specific area instead of being spread over the globe. This allows regional models to resolve explicitly smaller-scale meteorological phenomena that cannot be represented on the coarser grid of a global model. Regional models use a global model to specify conditions at the edge of their domain (boundary conditions) in order to allow systems from outside the regional model domain to move into its area. Uncertainty and errors within regional models are introduced by the global model used for the boundary conditions of the edge of the regional model, as well as errors attributable to the regional model itself.

Coordinate Systems

Horizontal Coordinates

Horizontal position may be expressed directly in geographic coordinates (latitude and longitude) for global models or in a map projection planar coordinates for regional models. The German weather service is using for its global ICON model(icosahedral non-hydrostatic global circulation model) a grid based on a regular icosahedron. Basic cells in this grid are triangles instead of the four corner cells in a traditional latitude-longitude grid. The advantage is that, different from a latitude-longitude cells are everywhere on the globe the same size. Disadvantage is that equations in this non rectangular grid are more complicated.

Vertical Coordinates

The vertical coordinate is handled in various ways. Lewis Fry Richardson's 1922 model used geometric height (z) as the vertical coordinate. Later models substituted the geometric coordinate with a pressure coordinate system, in which the geopotential heights of constant-pressure surfaces become dependent variables, greatly simplifying the primitive equations.This correlation between coordinate systems can be made since pressure decreases with height through the Earth's atmosphere. The first model used for operational forecasts, the single-layer barotropic model, used a single pressure coordinate at the 500-millibar (about 5,500 m (18,000 ft)) level, and thus was essentially two-dimensional. High-resolution models—also called *mesoscale models*—such as the Weather Research and Forecasting model tend to use normalized pressure coordinates referred to as sigma coordinates. This coordinate system receives its name from the independent variable used to scale atmospheric pressures with respect to the pressure at the surface, and in some cases also with the pressure at the top of the domain.

Model Output Statistics

Because forecast models based upon the equations for atmospheric dynamics do not perfectly determine weather conditions, statistical methods have been developed to attempt to correct the forecasts. Statistical models were created based upon the three-dimensional fields produced by numerical weather models, surface observations and the climatological conditions for specific locations. These statistical models are collectively referred to as model output statistics (MOS), and were developed by the National Weather Service for their suite of weather forecasting models in the late 1960s.

Model output statistics differ from the *perfect prog* technique, which assumes that the output of numerical weather prediction guidance is perfect. MOS can correct for local effects that cannot be resolved by the model due to insufficient grid resolution, as well as model biases. Because MOS is run after its respective global or regional model, its production is known as post-processing. Forecast parameters within MOS include maximum and minimum temperatures, percentage chance of rain within a several hour period, precipitation amount expected, chance that the precipitation will be frozen in nature, chance for thunderstorms, cloudiness, and surface winds.

Ensembles

In 1963, Edward Lorenz discovered the chaotic nature of the fluid dynamicsequations involved in weather forecasting. Extremely small errors in temperature, winds, or other initial inputs given to numerical models will amplify and double every five days, making it impossible for long-range forecasts—those made more than two weeks in advance—to predict the state of the atmosphere with any degree of forecast skill. Furthermore, existing observation networks have poor coverage in some regions (for example, over large bodies of water such as the Pacific Ocean), which introduces uncertainty into the true initial state of the atmosphere. While a set of equations, known as the Liouville equations, exists to determine the initial uncertainty in the model initialization, the equations are too complex to run in real-time, even with the use of supercomputers. These uncertainties limit forecast model accuracy to about five or six days into the future.

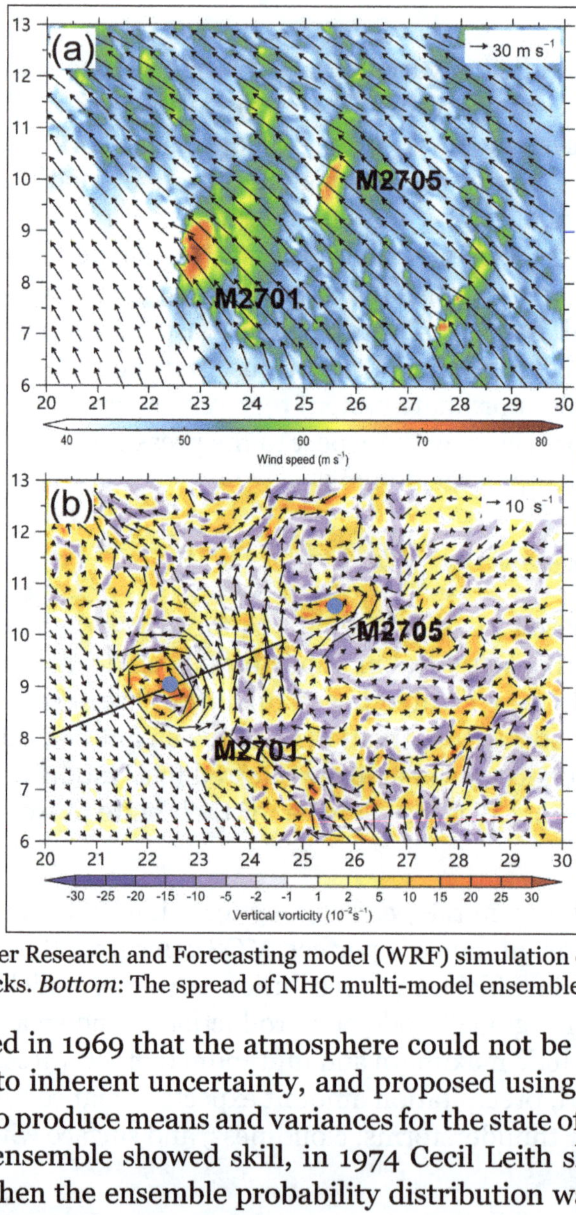

Top: Weather Research and Forecasting model (WRF) simulation of Hurricane Rita tracks. *Bottom*: The spread of NHC multi-model ensemble forecast.

Edward Epstein recognized in 1969 that the atmosphere could not be completely described with a single forecast run due to inherent uncertainty, and proposed using an ensemble of stochastic Monte Carlo simulations to produce means and variances for the state of the atmosphere. Although this early example of an ensemble showed skill, in 1974 Cecil Leith showed that they produced adequate forecasts only when the ensemble probability distribution was a representative sample of the probability distribution in the atmosphere.

Since the 1990s, *ensemble forecasts* have been used operationally (as routine forecasts) to account for the stochastic nature of weather processes – that is, to resolve their inherent uncertainty. This method involves analyzing multiple forecasts created with an individual forecast model by using different physical parametrizations or varying initial conditions. Starting in 1992 with ensemble forecasts prepared by the European Centre for Medium-Range Weather Forecasts (ECMWF) and the National Centers for Environmental Prediction, model ensemble forecasts have been used to help define the forecast uncertainty and to extend the window in which numerical weather forecasting is viable farther into the future than otherwise possible. The ECMWF model, the Ensemble Prediction System, uses singular vectors to simulate the initial probability density, while the NCEP ensemble, the Global Ensemble Forecasting System, uses a technique known as vector breeding. The UK Met Office runs global and regional ensemble forecasts where perturbations to initial conditions are produced using a Kalman filter. There are 24 ensemble members in the Met Office Global and Regional Ensemble Prediction System (MOGREPS).

In a single model-based approach, the ensemble forecast is usually evaluated in terms of an average of the individual forecasts concerning one forecast variable, as well as the degree of agreement between various forecasts within the ensemble system, as represented by their overall spread. Ensemble spread is diagnosed through tools such as spaghetti diagrams, which show the dispersion of one quantity on prognostic charts for specific time steps in the future. Another tool where ensemble spread is used is a meteogram, which shows the dispersion in the forecast of one quantity for one specific location. It is common for the ensemble spread to be too small to include the weather that actually occurs, which can lead to forecasters misdiagnosing model uncertainty; this problem becomes particularly severe for forecasts of the weather about ten days in advance.When ensemble spread is small and the forecast solutions are consistent within multiple model runs, forecasters perceive more confidence in the ensemble mean, and the forecast in general. Despite this perception, a *spread-skill relationship* is often weak or not found, as spread-error correlations are normally less than 0.6, and only under special circumstances range between 0.6–0.7. The relationship between ensemble spread and forecast skill varies substantially depending on such factors as the forecast model and the region for which the forecast is made.

In the same way that many forecasts from a single model can be used to form an ensemble, multiple models may also be combined to produce an ensemble forecast. This approach is called *multi-model ensemble forecasting*, and it has been shown to improve forecasts when compared to a single model-based approach. Models within a multi-model ensemble can be adjusted for their various biases, which is a process known as *superensemble forecasting*. This type of forecast significantly reduces errors in model output.

Applications

Air Quality Modeling

Air quality forecasting attempts to predict when the concentrations of pollutants will attain levels that are hazardous to public health. The concentration of pollutants in the atmosphere is determined by their *transport*, or mean velocity of movement through the atmosphere, their diffusion, chemical transformation, and ground deposition. In addition to pollutant source and terrain information, these models require data about the state of the fluid flow in the atmosphere to determine its transport and diffusion. Meteorological conditions such as thermal inversions can prevent surface air from rising, trapping pollutants near the surface, which makes accurate forecasts of such events crucial for air quality modeling. Urban air quality models require a very fine computational mesh, requiring the use of high-resolution mesoscale weather models; in spite of this, the quality of numerical weather guidance is the main uncertainty in air quality forecasts.

Climate Modeling

A General Circulation Model (GCM) is a mathematical model that can be used in computer simulations of the global circulation of a planetary atmosphere or ocean. An atmospheric general circulation model (AGCM) is essentially the same as a global numerical weather prediction model, and some (such as the one used in the UK Unified Model) can be configured for both short-term weather forecasts and longer-term climate predictions. Along with sea ice and land-surface components, AGCMs and oceanic GCMs (OGCM) are key components of global climate models, and are widely applied for understanding the climate and projecting climate change. For aspects of climate change, a range of man-made chemical emission scenarios can be fed into the climate models to see how an enhanced greenhouse effect would modify the Earth›s climate. Versions designed for climate applications with time scales of decades to centuries were originally created in 1969 by Syukuro Manabeand Kirk Bryan at the Geophysical Fluid Dynamics Laboratory in Princeton, New Jersey. When run for multiple decades, computational limitations mean that the models must use a coarse grid that leaves smaller-scale interactions unresolved.

Ocean Surface Modeling

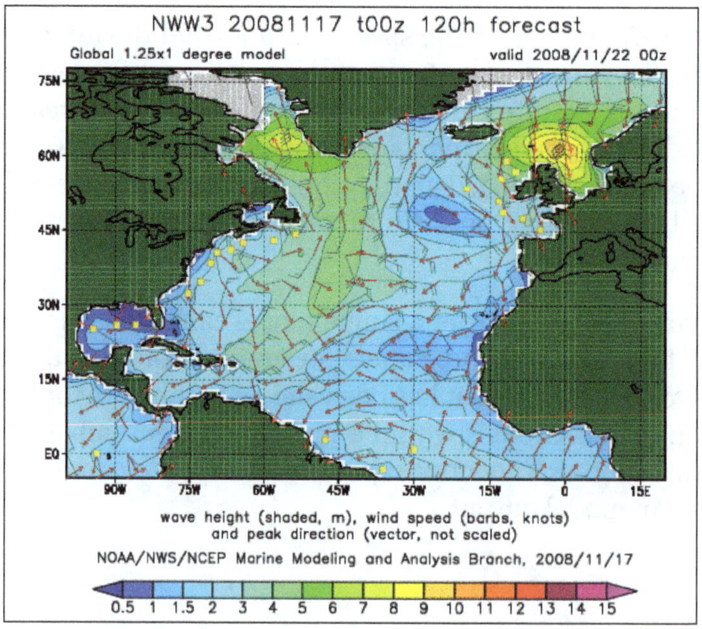

NOAA Wavewatch III 120-hour wind and wave forecast for the North Atlantic.

The transfer of energy between the wind blowing over the surface of an ocean and the ocean's upper layer is an important element in wave dynamics. The spectral wave transport equation is used to describe the change in wave spectrum over changing topography. It simulates wave generation, wave movement (propagation within a fluid), wave shoaling, refraction, energy transfer between waves, and wave dissipation. Since surface winds are the primary forcing mechanism in the spectral wave transport equation, ocean wave models use information produced by numerical weather prediction models as inputs to determine how much energy is transferred from the atmosphere into the layer at the surface of the ocean. Along with dissipation of energy through whitecaps and resonancebetween waves, surface winds from numerical weather models allow for more accurate predictions of the state of the sea surface.

Tropical Cyclone Forecasting

Tropical cyclone forecasting also relies on data provided by numerical weather models. Three main classes of tropical cyclone guidance models exist: Statistical models are based on an analysis of storm behavior using climatology, and correlate a storm›s position and date to produce a forecast that is not based on the physics of the atmosphere at the time. Dynamical models are numerical models that solve the governing equations of fluid flow in the atmosphere; they are based on the same principles as other limited-area numerical weather prediction models but may include special computational techniques such as refined spatial domains that move along with the cyclone. Models that use elements of both approaches are called statistical-dynamical models.

In 1978, the first hurricane-tracking model based on atmospheric dynamics—the movable fine-mesh (MFM) model—began operating. Within the field of tropical cyclone track forecasting, despite the ever-improving dynamical model guidance which occurred with increased computational power, it was not until the 1980s when numerical weather prediction showed skill, and until the 1990s when it consistently outperformed statistical or simple dynamical models. Predictions of the intensity of a tropical cyclone based on numerical weather prediction continue to be a challenge, since statistical methods continue to show higher skill over dynamical guidance.

Wildfire Modeling

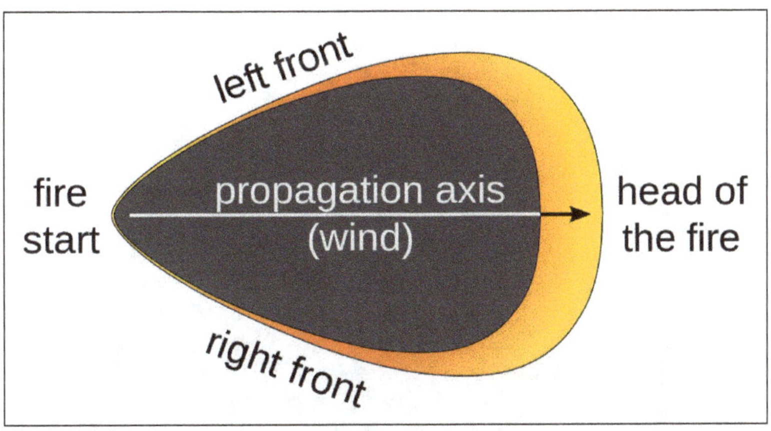

A simple wildfire propagation model.

On a molecular scale, there are two main competing reaction processes involved in the degradation of cellulose, or wood fuels, in wildfires. When there is a low amount of moisture in a cellulose fiber, volatilization of the fuel occurs; this process will generate intermediate gaseous products that will ultimately be the source of combustion. When moisture is present—or when enough heat is being carried away from the fiber, charring occurs. The chemical kinetics of both reactions indicate that there is a point at which the level of moisture is low enough—and/or heating rates high enough—for combustion processes become self-sufficient. Consequently, changes in wind speed, direction, moisture, temperature, or lapse rate at different levels of the atmosphere can have a significant impact on the behavior and growth of a wildfire. Since the wildfire acts as a heat source to the atmospheric flow, the wildfire can modify local advection patterns, introducing a feedback loop between the fire and the atmosphere.

A simplified two-dimensional model for the spread of wildfires that used convection to represent the effects of wind and terrain, as well as radiative heat transfer as the dominant method of heat transport led to reaction-diffusion systems of partial differential equations. More complex models join numerical weather models or computational fluid dynamics models with a wildfire component which allow the feedback effects between the fire and the atmosphere to be estimated. The additional complexity in the latter class of models translates to a corresponding increase in their computer power requirements. In fact, a full three-dimensional treatment of combustion via direct numerical simulation at scales relevant for atmospheric modeling is not currently practical because of the excessive computational cost such a simulation would require. Numerical weather models have limited forecast skill at spatial resolutions under 1 kilometer (0.6 mi), forcing complex wildfire models to parameterize the fire in order to calculate how the winds will be modified locally by the wildfire, and to use those modified winds to determine the rate at which the fire will spread locally. Although models such as Los Alamos' FIRETEC solve for the concentrations of fuel and oxygen, the computational grid cannot be fine enough to resolve the combustion reaction, so approximations must be made for the temperature distribution within each grid cell, as well as for the combustion reaction rates themselves.

NOWCASTING

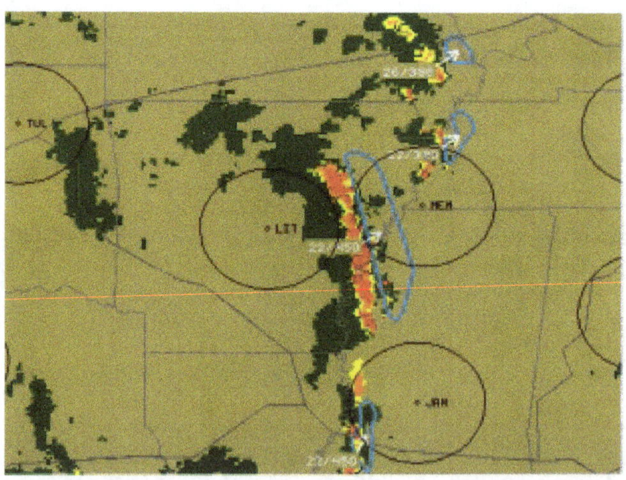

Forecast (blue lines) by " AutoNowcaster " for a thunderstorm line.

Nowcasting is weather forecasting on a very short term mesoscale period of up to 2 hours according to the World Meteorological Organization and up to six hours according to other authors in the field. This forecast is an extrapolation in time of known weather parameters, including those obtained by means of remote sensing, using techniques that take into account a possible evolution of the air mass. This type of forecast therefore includes details that cannot be solved by numerical weather prediction (NWP) models running over longer forecast periods.

Principle

Nowcasting in meteorology uses surface weather station data, wind profiler data, and any other weather data available to initialize the current weather situation and forecast by extrapolation for a period of 0 to 6 hours. In this time range it is possible to forecast small features such as individual storms with reasonable accuracy. Weather radar echoes and satellite data, giving cloud coverage, are particularly important in nowcasting because they are very detailed and pick out the size, shape, intensity, speed and direction of movement of individual features of weather on a continuous basis and a vastly better resolution than surface weather stations.

This used to be a simple extrapolation by a forecaster for the following few hours. But with the development of mesoscale numerical weather models, these information can be ingested into an expert system to produce a much better forecast combining numerical weather prediction and local effects not normally possible to be known beforehand. Different research groups, public and private, have developed such programs.

For instance, the French weather service, Météo-France, is using a software, named *ASPIC* to extrapolate to a fine scale the areas of precipitation. Other examples are *AutoNowcaster* which has been developed by UCAR to predict short term motion and evolution of thunderstorms, and private firms like ClimaCell using its proprietary HyperCast software for nowcasting precipitation type and intensity at 300-500 m geospatial resolution.

Usage

Data extrapolation, including development or dissipation, can be used to find the likely location of a moving weather system. The intensity of rainfall from a particular cloud or group of clouds can be estimated, giving a very good indication as to whether to expect flooding, the swelling of a river etc. Depending on the area of built-up space, drainage and land-use in general, a forecast warning may be issued.

Nowcasting is thus used for public safety, weather sensitive operation like snow removal, for aviation weather forecasts in both the terminal and en-route environment, marine safety, water and power management, off-shore oil drilling, construction industry and leisure industry. The strength of nowcasting lies in the fact that it provides location-specific forecasts of storm initiation, growth, movement and dissipation, which allows for specific preparation for a certain weather event by people in a specific location.

Research

The short term forecast is as old as weather forecasting itself. During the nineteenth century, the first modern meteorologists were using extrapolation methods for predicting the movement of low pressure systems and anticyclones on surface maps. The researchers subsequently applied the laws of fluid dynamics to the atmosphere and developed the NWP as we know it today. However, the data resolution and parameterization of meteorological primitive equations still leave uncertainty about the small-scale projections, in time and space.

The arrival of remote sensing means, such as radar and satellite, and more rapid development of the computer, greatly help to fill that gap. For instance, digital radar systems made it possible to track thunderstorms, providing users with the ability to acquire detailed information of each storm tracked, since the late 1980's. They are first identified by matching precipitation raw data to a set of preprogrammed characteristics into the system, including signs of organization in the horizontal and continuity in the vertical. Once the thunderstorm cell is identified, speed, distance covered, direction, and Estimated Time of Arrival (ETA) are all tracked and recorded to be utilized later.

In 2017, the arrival of passive sensing means, such as wireless networks, helped progress nowcasting even further. It became possible to receive inputs every minute and achieve greater accuracy in short-term forecasting.

Several countries have developed nowcasting programs as previously mentioned. The World Meteorological Organization (WMO) supports these efforts and held test campaigns of such systems at various occasions. For example, during the Olympic Games in Sydney and Beijing, several countries were invited to use their software to support the Games.

Several scientific conferences addressing the topic. In 2009, WMO has even organized a symposium devoted to Nowcasting.

TERMINAL AERODROME FORECAST

A terminal aerodrome forecast is a report established for the 5 statute mile radius around an airport. TAF reports are usually given for larger airports. Each TAF is valid for a 24-hour time period, and is updated four times a day at 0000Z, 0600Z, 1200Z, and 1800Z. The TAF utilizes the same descriptors and abbreviations as used in the METAR report.

The terminal forecast includes the following information in sequential order:

- Type of Report A TAF can be either a routine forecast (TAF) or an amended forecast (TAF AMD).
- ICAO Station Identifier The station identifier is the same as that used in a METAR.
- Date and Time of Origin Time and date of TAF origination is given in the six-number code with the first two being the date, the last four being the time. Time is always given in UTC as denoted by the Z following the number group.
- Valid Period Date and Time The valid forecast time period is given by a six-digit number group. The first two numbers indicate the date, followed by the two-digit beginning time for the valid period, and the last two digits are the ending time.
- Forecast Wind The wind direction and speed forecast are given in a five-digit number group. The first three indicate the direction of the wind in reference to true north. The last two digits state the windspeed in knots as denoted by the letters "KT." Like the METAR, winds greater than 99 knots are given in three digits.
- Forecast Visibility The forecast visibility is given in statute miles and may be in whole numbers or fractions. If the forecast is greater than 6 miles, it will be coded as "P6SM."

- Forecast Significant Weather Weather phenomenon is coded in the TAF reports in the same format as the METAR. If no significant weather is expected during the forecast time period, the denotation "NSW" will be included in the "becoming" or "temporary" weather groups.
- Forecast Sky Condition Forecast sky con- ditions are given in the same manner as the METAR. Only cumulonimbus (CB) clouds are forecast in this portion of the TAF report as opposed to CBs and towering cumulus in the METAR.
- Forecast Change Group For any significant weather change forecast to occur during the TAF time period, the expected conditions and time period are included in this group. This information may be shown as From (FM), Becoming (BECMG), and Temporary (TEMPO). "From" is used when a rapid and significant change, usually within an hour, is expected. "Becoming" is used when a gradual change in the weather is expected over a period of no more than 2 hours. "Temporary" is used for temporary fluctuations of weather, expected to last for less than an hour.
- Probability Forecast The probability forecast is given percentage that describes the probability of thunderstorms and precipitation occurring in the coming hours. This forecast is not used for the first 6 hours of the 24-hour forecast.

The period of validity should not be less than 6 HRs, no more than 30 HRs.

TAFs valid for less than 12 HRs should be issued every 3 HRs and those valid for 12 to 30 HRs every 6 HRs.

Application of Initial Part of TAF

- Applicable time period: From the start of the TAF validity period up to the time of applicability of the first subsequent "FM" or "BECMG" or, if no "FM" or "BECMG" is given, up to the end of the validity period of the TAF.
- Application of forecast: The prevailing weather conditions forecast in the initial part of the TAF should be fully applied with the exception of the mean wind and gusts (and crosswind) which should be applied in accordance with the policy in the column "BECMG AT" and "FM" in the table on the next page. This may, however, be overruled temporarily by a "TEMPO" or "PROB" if applicable, according to the table.

Change indicator	Explanation
BECMG (becoming)	Changes where the meteorological conditions are expected to reach or pass through specified threshold values at a regular or irregular rate and at an unspecified time during the time period (starting 2300Z on the 8th day and ending at 0100Z on the 9th day). Example TAF – FT: NZZO 081140Z 0812/0918 VRB02KT 1800 BR BECMG 0823/0901 00000KT 0500 FG VV002
TEMPO (temporary)	Describes expected frequent or infrequent temporary fluctuations in meteorological conditions which reach or pass specified threshold values and last for a period of less then 1 hour in each instance and in the aggregate, cover less than half of the fluctuation period (from 1800Z on the 8th day to 2300Z on the 8th day). Example TAF – FC: LFPG 081140Z 0812/0918 29010KT 5000 NSW TEMPO 0818/0823 1500 SHSN =

PROB (probability)	Probability in percent of occurrence of an alternative weather development and the time period (from 1000Z on the 9th day to 1200Z on the 9th day). PROB30/40 alone or in combination with TEMPO are used. Example TAF – FT 18/24: VAAB 081140Z 0812/0918 26018KT 3000 RA SCT012 BKN030 PROB40 0910/0912 25028G40+RA =
Time Divider FM (from)	Where one set of prevailing weather condition is expected to change significantly and more or less completely to a different set of conditions, the period of validity should be subdivided into a self-contained period giving the time group in whole hour and minutes UTC when the change is expected to occur (at 1000Z on the 9th day). The period following from should be self-contained and all forecast conditions given before should be superseded. Example TAF – FT: KBGR 081140Z 0812/0918 35008KT P6SM SCT200 FM091000 03005KT P6SM BKN120 =

	Optional Groups in TAF
Turbulence	Coded 6-figure group 1st figure: Group indicator 5 2nd figure: Type of turbulence 0 = none 1 = light 2 = moderate in clear air, infrequent 3 = moderate in clear air, frequent 4 = moderate in cloud, infrequent 5 = moderate in cloud, frequent 6 = severe in clear air, infrequent 7 = severe in clear air, frequent 8 = severe in cloud, infrequent 9 = severe in cloud, frequent X = extreme (US AFB TAF only) 3rd to 5th figure: Height of lowest turbulence level in hundreds of feet above aerodrome elevation. 6th figure: Thickness of turbulent layer in thousands of feet; exception: 0 = top of clouds Example: 590200 = severe in cloud, frequent, from 2000ft up to top of cloud.
Icing	Coded 6-figure group 1st figure: Group indicator 6 2nd figure: Type of icing 0 = none 1 = light 2 = light in cloud 3 = light in precipitation 4 = moderate 5 = moderate in cloud

Icing	6 = moderate in precipitation
	7 = severe
	8 = severe in cloud
	9 = severe in precipitation
	3rd to 5th figure: Height of lowest icing level in hundreds of feet above aerodrome elevation.
	6th figure: Thickness of icing layer in thousands of feet; exception: 0 = top of cloud
	Example: 660083 = moderate in precipitation, from 800 upwards in layer 3000ft thick.

MARINE WEATHER FORECASTING

Marine weather forecasting is the process by which marinersand meteorological organizations attempt to forecast future weather conditions over the Earth's oceans. Mariners have had rules of thumb regarding the navigation around tropical cyclonesfor many years, dividing a storm into halves and sailing through the normally weaker and more navigable half of their circulation. Marine weather forecasts by various weather organizations can be traced back to the sinking of the Royal Charter in 1859 and the RMS Titanic in 1912.

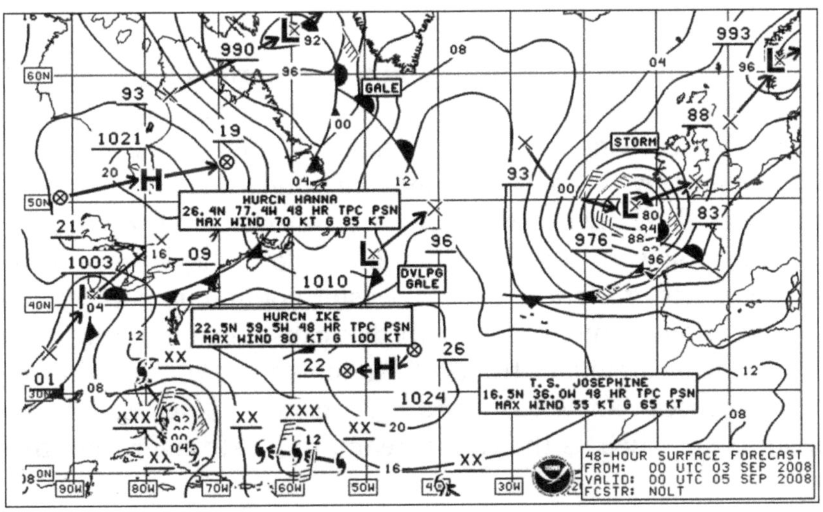

OPC pressure forecast valid at 48 hours.

The wind is the driving force of weather at sea, as wind generates local wind waves, long ocean swells, and its flow around the subtropical ridge helps maintain warm water currents such as the Gulf Stream. The importance of weather over the ocean during World War II led to delayed or secret weather reports, in order to maintain a competitive advantage. Weather ships were established by various nations during World War II for forecasting purposes, and were maintained through 1985 to help with transoceanic plane navigation.

Voluntary observations from ships, weather buoys, weather satellites, and numerical weather prediction have been used to diagnose and help forecast weather over the Earth›s ocean areas. Since the 1960s, numerical weather prediction's role over the Earth's seas has taken a greater role in the forecast process. Weather elements such as sea state, surface winds, tide levels, and sea surface temperature are tackled by organizations tasked with forecasting weather over open oceans and seas. Currently, the Japan Meteorological Agency, the United States National Weather Service, and the United Kingdom Met Office create marine weather forecasts for the Northern Hemisphere.

Importance of the Wind

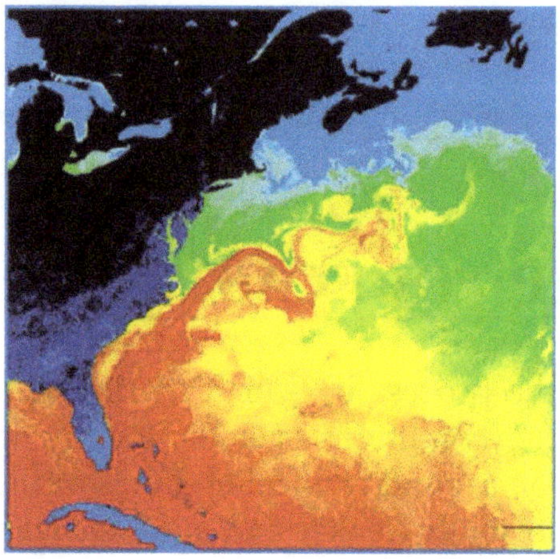

Surface temperature in the western North Atlantic, the Gulf Stream is in red.

Development of Warm Ocean Currents

The trade winds blow westward in the tropics, and the westerliesblow eastward at mid-latitudes. This wind pattern applies a stress to the subtropical ocean surface with negative curl across the north Atlantic Ocean. The resulting Sverdrup transport is equatorward.Because of conservation of potential vorticity caused by the poleward-moving winds on the subtropical ridge's western periphery and the increased relative vorticity of northward moving water, transport is balanced by a narrow, accelerating poleward current, which flows along the western boundary of the ocean basin, outweighing the effects of friction with the western boundary current known as the Labrador current. The conservation of potential vorticity also causes bends along the Gulf Stream, which occasionally break off due to a shift in the Gulf Stream's position, forming separate warm and cold eddies.This overall process, known as western intensification, causes currents on the western boundary of an ocean basin, such as the Gulf Stream, to be stronger than those on the eastern boundary.

Swell Dispersion and Wave Groups

North Pacific storm waves as seen from the NOAA M/V *Noble Star*.

Swells are often created by storms long distances away from the beach where they break, and the propagation of the longest swells is only limited by shorelines. For example, swells generated in the Indian Ocean have been recorded in California after more than half a round-the-world trip. This distance allows the waves comprising the swells to be better sorted and free of chop as they travel toward the coast. Waves generated by storm winds have the same speed and will group together and travel with each other, while others moving at even a fraction of a metre per second slower will lag behind, ultimately arriving many hours later due to the distance covered. The time of propagation from the source *t* is proportional to the distance *X* divided by the wave period *T*. In deep water it is $t = 4\pi X / (gT)$ where g is the acceleration of gravity. As an example, for a storm located 10,000 kilometres (6,200 mi) away, swells with a period *T*=15 s will arrive 10 days after the storm, followed by 14 s swells another 17 hours later.

This dispersive arrivals of swells, long periods first with a reduction in the peak wave period over time, can be used to tell the distance at which swells were generated. Whereas the sea state in the storm has a frequency spectrum with more or less always the same shape (i.e. a well defined peak with dominant frequencies within plus or minus 7% of the peak), the swell spectra are more and more narrow, sometimes as 2% or less, as waves disperse further and further away. The result is that wave groups (called sets by surfers) can have a large number of waves. From about seven waves per group in the storm, this rises to 20 and more in swells from very distant storms.

Sailing Ship Journeys

Ocean journeys by sailing ship can take many months, and a common hazard is becoming becalmed because of lack of wind, or being blown off course by severe storms or winds that do not allow progress in the desired direction. A severe storm could lead to shipwreck, and the loss of all hands. Sailing ships can only carry a certain quantity of supplies in their hold, so they have to plan long voyages carefully to include appropriate provisions, including fresh water.

Tropical Cyclone Avoidance

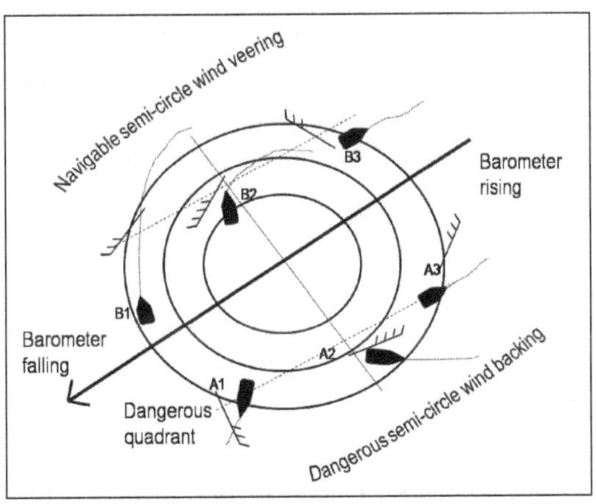

The dangerous semicircle is the upper-right corner, with the arrow marking the direction of motion of a Northern Hemisphere storm.

Mariners have a way to safely navigate around tropical cyclones. They split tropical cyclones in two, based on their direction of motion, and maneuver to avoid the right segment of the cyclone in the Northern Hemisphere (the left in the Southern Hemisphere). Sailors term the right side the **dangerous semicircle** since the heaviest rain and strongest winds and seas were located in this half of the storm, as the cyclone›s translation speed and its rotational wind are additive. The other half of the tropical cyclone is called the **navigable semicircle**since weather conditions are lessened (subtractive) in this portion of the storm. The rules of thumb for ship travel when a tropical cyclone is in their vicinity are to avoid them if at all possible and do not cross their forecast path (crossing the T). Those traveling through the dangerous semicircle are advised to keep to the true wind on the starboard bow and make as much headway as possible. Ships moving through the navigable semicircle are advised to keep the true wind on the starboard quarter while making as much headway as possible.

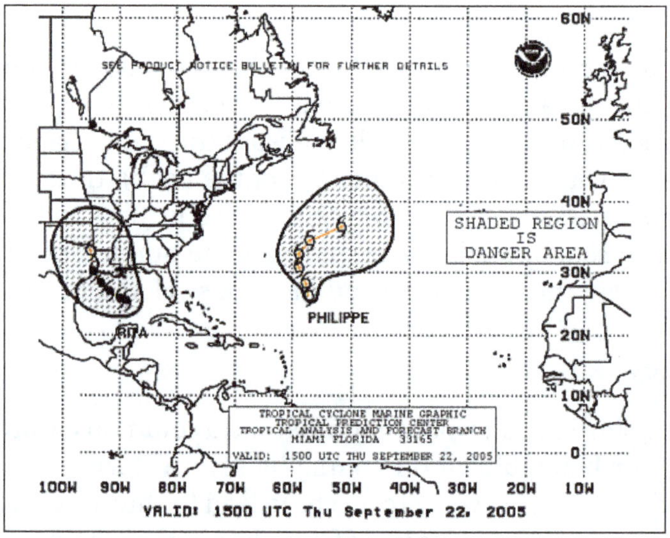

Hurricanes Rita and Philippe shown with 1-2-3 rule predictions.

The 1-2-3 rule (mariners' 1-2-3 rule or danger area) is a guideline commonly taught to mariners for severe storm (specifically hurricane and tropical storm) tracking and prediction. It refers to the rounded long-term National Hurricane Center forecast errors of 100-200-300 nautical miles at 24-48-72 hours, respectively. However, these errors have decreased to near 50-100-150 as NHC forecasters become more accurate with tropical cyclone track forecasting. The "danger area" to be avoided is constructed by expanding the forecast path by a radius equal to the respective hundreds of miles plus the forecast wind radii (size of the storm at those hours).

Within Numerical Weather Prediction

Ocean Surface Modeling

The transfer of energy between the wind blowing over the surface of an ocean and the ocean's upper layer is an important element in wave dynamics. The spectral wave transport equation is used to describe the change in wave spectrum over changing topography. It simulates wave generation, wave movement (propagation within a fluid), wave shoaling, refraction, energy transfer between waves, and wave dissipation. Since surface winds are the primary forcing mechanism in the spectral wave transport equation, ocean wave models use information produced by numerical weather prediction models as inputs to determine how much energy is transferred from the atmosphere into the layer at the surface of the ocean. Along with dissipation of energy through whitecaps and resonance between waves, surface winds from numerical weather models allow for more accurate predictions of the state of the sea surface.

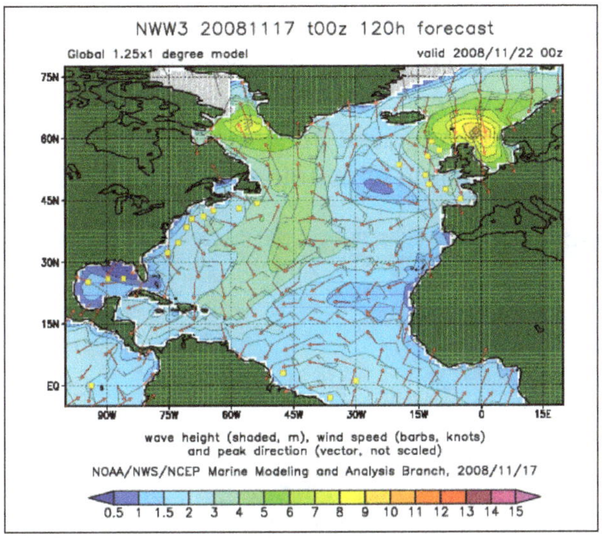

NOAA Wavewatch III 120-hour wind and wave forecast for the North Atlantic.

The first ocean wave models were developed in the 1960s and 1970s. These models had the tendency to overestimate the role of wind in wave development and underplayed wave interactions. A lack of knowledge concerning how waves interacted among each other, assumptions regarding a maximum wave height, and deficiencies in computer power limited the performance of the models. After experiments were performed in 1968, 1969, and 1973, wind input from the Earth›s atmosphere was weighted more accurately in the predictions. A second generation of models was developed in the 1980s, but they could not realistically model swell nor depict wind-driven waves (also known as wind waves) caused by rapidly changing wind fields, such as those within tropical cyclones. This caused the development of a third generation of wave models from 1988 onward.

Within this third generation of models, the spectral wave transport equation is used to describe the change in wave spectrum over changing topography. It simulates wave generation, wave movement (propagation within a fluid), wave shoaling, refraction, energy transfer between waves, and wave dissipation. Since surface winds are the primary forcing mechanism in the spectral wave transport equation, ocean wave models use information produced by numerical weather prediction models as inputs to determine how much energy is transferred from the atmosphere into the layer at the surface of the ocean. Along with dissipation of energy through whitecaps and resonance between waves, surface winds from numerical weather models allow for more accurate predictions of the state of the sea surface.

Observing Platforms

Weather Ships

The idea of a stationary weather ship was proposed as early as 1921 by Météo-France to help support shipping and the coming of transatlantic aviation. Established during World War II, a weather ship, or ocean weather vessel, was a ship stationed in the ocean as a platform for surface and upper air meteorological observations for use in weather forecasting. They were used during World War II but had no means of defense, which led to the loss of several ships and many lives. They were primarily located in the north Atlantic and north Pacific oceans, reporting via radio. In addition to their weather reporting function, these vessels aided in search and rescue operations, supported transatlantic flights, acted as research platforms for oceanographers, monitored marine pollution, and aided weather forecasting both by weather forecasters and within computerized atmospheric models. Research vessels remain heavily used in oceanography, including physical oceanography and the integration of meteorological and climatological data in Earth system science.

The weather ship MS *Polarfront* at sea.

The establishment of weather ships proved to be so useful during World War II that the International Civil Aviation Organization (ICAO) had established a global network of 13 weather ships by 1948, with seven operated by the United States, one operated jointly by the United States and Canada, two supplied by the United Kingdom, one maintained by France, one a joint venture by the Netherlands and Belgium, and one shared by the United Kingdom, Norway, and Sweden. This number was eventually negotiated down to nine. The agreement of the use of weather ships by the international community ended in 1985.

Weather Buoys

Weather buoys are instruments which collect weather and ocean data within the world›s oceans, as well as aid during emergency response to chemical spills, legal proceedings, and engineering design. Moored buoys have been in use since 1951, while drifting buoys have been used since 1972. Moored buoys are connected with the ocean bottom using either chains, nylon, or buoyant polypropylene. With the decline of the weather ship, they have taken a more primary role in measuring conditions over the open seas since the 1970s. During the 1980s and 1990s, a network of buoys in the central and eastern tropical Pacific Ocean helped study the El Niño-Southern Oscillation. Moored weather buoys range from 1.5 metres (4.9 ft) to 12 metres (39 ft) in diameter, while drifting buoys are smaller, with diameters of 30 centimetres (12 in) to 40 centimetres (16 in).Drifting buoys are the dominant form of weather buoy in sheer number, with 1250 located worldwide. Wind data from buoys has smaller error than that from ships. There are differences in the values of sea surface temperature measurements between the two platforms as well, relating to the depth of the measurement and whether or not the water is heated by the ship which measures the quantity.

Weather buoy operated by the NOAA National Data Buoy Center.

Weather Satellites

The first image obtained from the GOES 1 satellite.

In use since 1960, the weather satellite is a type of satellite that is primarily used to monitor the weather and climate of the Earth. Satellites can be polar orbiting, covering the entire Earth asynchronously, or geostationary, hovering over the same spot on the equator. Meteorological satellites see more than clouds and cloud systems. Beginning with the Nimbus 3 satellite in 1969, temperature information through the atmospheric column began to be retrieved by satellites from the eastern Atlantic and most of the Pacific Ocean, which led to significant forecast improvements. City lights, fires, effects of pollution, auroras, sand and dust storms, snow cover, ice mapping, boundaries of ocean currents, energy flows, etc., and other types of environmental information are collected using weather satellites. Other environmental satellites can detect changes in the Earth›s vegetation, sea state, ocean color, and ice fields. El Niño and its effects on weather are monitored daily from satellite images. Collectively, weather satellites flown by the U.S., Europe, India, China, Russia, and Japan provide nearly continuous observations for a global weather watch.

Utility

Commercial and recreational use of waterways can be limited significantly by wind direction and speed, waveperiodicity and heights, tides, and precipitation. These factors can each influence the safety of marine transit. Consequently, a variety of codes have been established to efficiently transmit detailed marine weather forecasts to vessel pilots via radio, for example the MAFOR (marine forecast). Typical weather forecasts can be received at sea through the use of RTTY, Navtex and Radiofax.

NCEP Products Available

Marine weather warnings and forecasts in print and prognostic chart formats are produced for up five days into the future. Forecasts in printed form include the High Seas Forecast, Offshore Marine Forecasts, and Coastal Waters Forecasts. To help shorten the length of the forecast products, single words and phrases are used to describe areas out at sea. Experimental gridded significant wave height forecasts began being produced by the Ocean Prediction Center in 2006, a first step toward digital marine service for high seas and offshore areas. Additional gridded products such as surface pressure and winds are under development. Recently, National Weather Service operational extratropical storm surge model output to provide experimental extratropical storm surge guidance for coastal weather forecast offices to assist them in coastal flood warning and forecast operations.

Responsible Organizations and their Areas

Northern Hemisphere

Within the Japan Meteorological Agency, marine observatories are seated in Hakodate, Maizuru, Kobe and Nagasaki. These stations observe ocean waves, tide levels, sea surface temperature and ocean current etc. in the Northwestern Pacific basin, as well as the Sea of Japan and the Sea of Okhotsk basin, and provide marine meteorological forecasts resulted from them, in cooperation with the Hydrographic and Oceanographic Department, Japan Coast Guard.

Within the United Kingdom, the Shipping Forecast is a BBC Radio broadcast of weather reports and forecasts for the seas around the coasts of the British Isles. It is produced by the Met Office and broadcast four times per day by BBC Radio 4 on behalf of the Maritime and Coastguard Agency. The forecasts sent over the Navtexsystem use a similar format and the same sea areas. The waters around the British Isles are divided into sea areas, also known as weather areas.

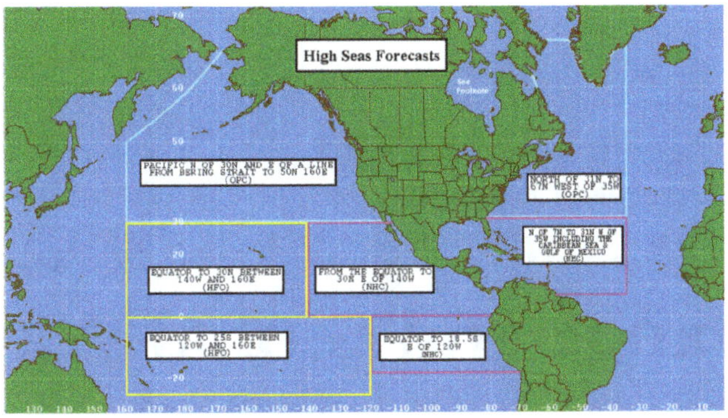

The National Weather Service areas of marine weather forecasting responsibility.

Within the United States National Weather Service, the Ocean Prediction Center (OPC), established in 1995, is one of the National Centers for Environmental Prediction's (NCEP's) original six service centers. Until January 12, 2003, the name of the organization was the Marine Prediction Center. The OPC issues forecasts up to five days in advance for ocean areas north of 31 north latitude and west of 35 west longitude in the Atlantic, and across the northeast Pacific north of 30 north latitude and east of 160 east longitude. Until recently, the OPC provided forecast points for tropical cyclones north of 20 north latitude and east of the 60 west longitude to the National Hurricane Center. OPC is composed of two branches: the Ocean Forecast Branch and the Ocean Applications Branch. The National Hurricane Center covers marine areas south of the 31st parallel in the Atlantic and 30th parallel in the Pacific between the 35th meridian west and 140th meridian west longitude. The Honolulu Weather Service Forecast Office forecasts within the area between the 140th meridian west and the 160th meridian east, from the 30th parallel north down to equator.

Southern Hemisphere

The National Hurricane Center's area of responsibility includes Southern Hemisphere areas in the Pacific down to 18.5 degrees south eastward of the 120th meridian west. South of the equation, the NWS Honolulu Forecast Office forecasts southward to the 25th parallel south between the 160th meridian east and the 120th meridian west.

References

- Weather-forecasting, science: britannica.com, Retrieved 29 April, 2019
- Lynch, peter (2006). "weather prediction by numerical process". The emergence of numerical weather prediction. Cambridge university press. Pp. 1–27. Isbn 978-0-521-85729-1
- Translation bureau. "nowcasting". Termium plus. Public works and government services canada. Retrieved may 12, 2016
- Terminal-aerodrome-forecast, meteorology: flightcrewguide.com, Retrieved 16 June, 2019
- Glossary of Meteorology (2009). "trade winds". Glossary of Meteorology. American Meteorological Society. Archived from the original on 2008-12-11. Retrieved 2008-09-08
- National Data Buoy Center (2008-02-04). "Moored Buoy Program". National Oceanic and Atmospheric Administration. Archived from the originalon 2011-01-03. Retrieved 2011-01-29

7

Climate Change and Global Warming

Climate change is a condition when the changes in Earth's climate system cause new weather patterns. The long term increase in the average temperature of the Earth's climate system is termed as global warming. The chapter closely examines the key concepts related to climate change such as ozone depletion and global warming to provide an extensive understanding of the subject.

CLIMATE CHANGE

Climate change is the periodic modification of Earth's climate brought about as a result of changes in the atmosphere as well as interactions between the atmosphere and various other geologic, chemical, biological, and geographic factors within the Earth system.

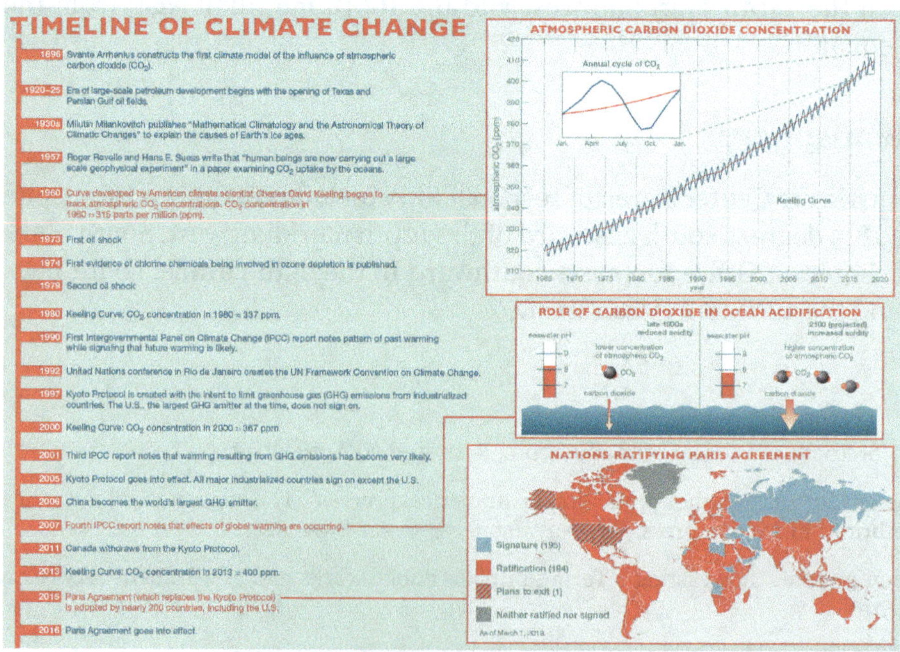

Climate change: timeline.

The atmosphere is a dynamic fluid that is continually in motion. Both its physical properties and its rate and direction of motion are influenced by a variety of factors, including solar radiation, the geographic position of continents, ocean currents, the location and orientation of mountain ranges, atmospheric chemistry, and vegetation growing on the land surface. All these factors change through time. Some factors, such as the distribution of heat within the oceans, atmospheric chemistry, and surface vegetation, change at very short timescales. Others, such as the position of continents and the location and height of mountain ranges, change over very long timescales. Therefore, climate, which results from the physical properties and motion of the atmosphere, varies at every conceivable timescale.

Climate is often defined loosely as the average weather at a particular place, incorporating such features as temperature, precipitation, humidity, and windiness. A more specific definition would state that climate is the mean state and variability of these features over some extended time period. Both definitions acknowledge that the weather is always changing, owing to instabilities in the atmosphere. And as weather varies from day to day, so too does climate vary, from daily day-and-night cycles up to periods of geologic time hundreds of millions of years long. In a very real sense, climate variation is a redundant expression—climate is always varying. No two years are exactly alike, nor are any two decades, any two centuries, or any two millennia.

The Earth System

Iceberg: Tourist boat in front of a massive iceberg near the coast of Greenland.

The atmosphere is influenced by and linked to other features of Earth, including oceans, ice masses (glaciers and sea ice), land surfaces, and vegetation. Together, they make up an integrated Earth system, in which all components interact with and influence one another in often complex ways. For instance, climate influences the distribution of vegetation on Earth's surface (e.g., deserts exist in arid regions, forests in humid regions), but vegetation in turn influences climate by reflecting radiant energy back into the atmosphere, transferring water (and latent heat) from soil to the atmosphere, and influencing the horizontal movement of air across the land surface.

Earth scientists and atmospheric scientists are still seeking a full understanding of the complex feedbacks and interactions among the various components of the Earth system. This effort is being facilitated by the development of an interdisciplinary science called Earth system science. Earth system science is composed of a wide range of disciplines, including climatology (the study of the atmosphere), geology (the study of Earth's surface and underground processes), ecology (the study of how Earth's organisms relate to one another and their environment), oceanography (the study of Earth's oceans), glaciology (the study of Earth's ice masses), and even the social sciences (the study of human behaviour in its social and cultural aspects).

Drought-resistant plants in the Repetek Preserve in the southeastern Karakum Desert, Turkmenistan.

A full understanding of the Earth system requires knowledge of how the system and its components have changed through time. The pursuit of this understanding has led to development of Earth system history, an interdisciplinary science that includes not only the contributions of Earth system scientists but also paleontologists (who study the life of past geologic periods), paleoclimatologists (who study past climates), paleoecologists (who study past environments and ecosystems), paleoceanographers (who study the history of the oceans), and other scientists concerned with Earth history. Because different components of the Earth system change at different rates and are relevant at different timescales, Earth system history is a diverse and complex science. Students of Earth system history are not just concerned with documenting what has happened; they also view the past as a series of experiments in which solar radiation, ocean currents, continental configurations, atmospheric chemistry, and other important features have varied. These experiments provide opportunities to learn the relative influences of and interactions between various components of the Earth system. Studies of Earth system history also specify the full array of states the system has experienced in the past and those the system is capable of experiencing in the future.

Deciduous forest in fall coloration, Wasatch Mountains, Utah.

Undoubtedly, people have always been aware of climatic variation at the relatively short timescales of seasons, years, and decades. Biblical scripture and other early documents refer to droughts, floods, periods of severe cold, and other climatic events. Nevertheless, a full appreciation of the nature and magnitude of climatic change did not come about until the late 18th and early 19th centuries, a time when the widespread recognition of the deep antiquity of Earth occurred. Naturalists of this time, including Scottish geologist Charles Lyell, Swiss-born naturalist and geologist Louis Agassiz, English naturalist Charles Darwin, American botanist Asa Gray, and Welsh naturalist Alfred Russel Wallace, came to recognize geologic and biogeographic evidence that made sense only in the light of past climates radically different from those prevailing today.

Geologists and paleontologists in the 19th and early 20th centuries uncovered evidence of massive climatic changes taking place before the Pleistocene—that is, before some 2.6 million years ago. For example, red beds indicated aridity in regions that are now humid (e.g., England and New England), whereas fossils of coal-swamp plants and reef corals indicated that tropical climates once occurred at present-day high latitudes in both Europe and North America. Since the late 20th century the development of advanced technologies for dating rocks, together with geochemical techniques and other analytical tools, have revolutionized the understanding of early Earth system history.

The occurrence of multiple epochs in recent Earth history during which continental glaciers, developed at high latitudes, penetrated into northern Europe and eastern North America was recognized by scientists by the late 19th century. Scottish geologist James Croll proposed that recurring variations in orbital eccentricity (the deviation of Earth's orbit from a perfectly circular path) were responsible for alternating glacial and interglacial periods. Croll's controversial idea was taken up by Serbian mathematician and astronomer Milutin Milankovitch in the early 20th century. Milankovitch proposed that the mechanism that brought about periods of glaciation was driven by cyclic changes in eccentricity as well as two other orbital parameters: precession (a change in the directional focus of Earth's axis of rotation) and axial tilt (a change in the inclination of Earth's axis with respect to the plane of its orbit around the Sun). Orbital variation is now recognized as an important driver of climatic variation throughout Earth's history.

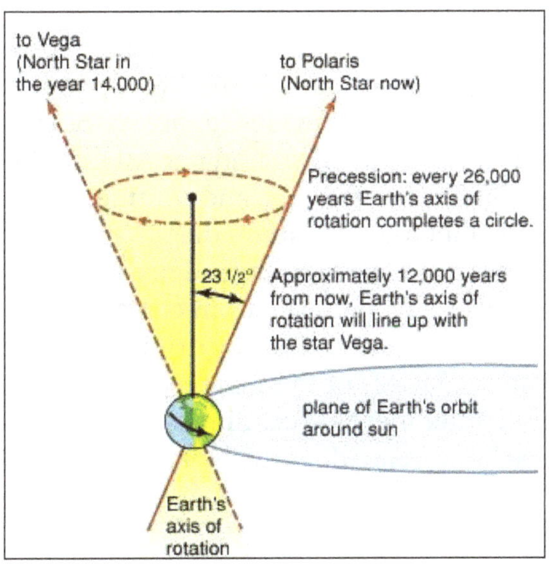

The precession of Earth's axis.

Evidence for Climate Change

All historical sciences share a problem: As they probe farther back in time, they become more reliant on fragmentary and indirect evidence. Earth system history is no exception. High-quality instrumental records spanning the past century exist for most parts of the world, but the records become sparse in the 19th century, and few records predate the late 18th century. Other historical documents, including ship's logs, diaries, court and church records, and tax rolls, can sometimes be used. Within strict geographic contexts, these sources can provide information on frosts, droughts, floods, sea ice, the dates of monsoons, and other climatic features—in some cases up to several hundred years ago.

Fortunately, climatic change also leaves a variety of signatures in the natural world. Climate influences the growth of trees and corals, the abundance and geographic distribution of plant and animal species, the chemistry of oceans and lakes, the accumulation of ice in cold regions, and the erosion and deposition of materials on Earth's surface. Paleoclimatologists study the traces of these effects, devising clever and subtle ways to obtain information about past climates. Most of the evidence of past climatic change is circumstantial, so paleoclimatology involves a great deal of investigative work. Wherever possible, paleoclimatologists try to use multiple lines of evidence to cross-check their conclusions. They are frequently confronted with conflicting evidence, but this, as in other sciences, usually leads to an enhanced understanding of the Earth system and its complex history. New sources of data, analytical tools, and instruments are becoming available, and the field is moving quickly. Revolutionary changes in the understanding of Earth's climate history have occurred since the 1990s, and coming decades will bring many new insights and interpretations.

Ongoing climatic changes are being monitored by networks of sensors in space, on the land surface, and both on and below the surface of the world's oceans. Climatic changes of the past 200–300 years, especially since the early 1900s, are documented by instrumental records and other archives. These written documents and records provide information about climate change in some locations for the past few hundred years. Some very rare records date back over 1,000 years. Researchers studying climatic changes predating the instrumental record rely increasingly on natural archives, which are biological or geologic processes that record some aspect of past climate. These natural archives, often referred to as proxy evidence, are extraordinarily diverse; they include, but are not limited to, fossil records of past plant and animal distributions, sedimentary and geochemical indicators of former conditions of oceans and continents, and land surface features characteristic of past climates. Paleoclimatologists study these natural archives by collecting cores, or cylindrical samples, of sediments from lakes, bogs, and oceans; by studying surface features and geological strata; by examining tree ring patterns from cores or sections of living and dead trees; by drilling into marine corals and cave stalagmites; by drilling into the ice sheets of Antarctica and Greenland and the high-elevation glaciers of the Plateau of Tibet, the Andes, and other montane regions; and by a wide variety of other means. Techniques for extracting paleoclimatic information are continually being developed and refined, and new kinds of natural archives are being recognized and exploited.

Causes of Climate Change

It is much easier to document the evidence of climate variability and past climate change than it is to determine their underlying mechanisms. Climate is influenced by a multitude of factors that operate at timescales ranging from hours to hundreds of millions of years. Many of the causes of climate change are external to the Earth system. Others are part of the Earth system but external to the atmosphere. Still others involve interactions between the atmosphere and other components of the Earth system and are collectively described as feedbacks within the Earth system. Feedbacks are among the most recently discovered and challenging causal factors to study. Nevertheless, these factors are increasingly recognized as playing fundamental roles in climate variation.

Solar Variability

The luminosity, or brightness, of the Sun has been increasing steadily since its formation. This phenomenon is important to Earth's climate, because the Sun provides the energy to drive atmospheric circulation and constitutes the input for Earth's heat budget. Low solar luminosity during Precambrian time underlies the faint young Sun paradox.

Radiative energy from the Sun is variable at very small timescales, owing to solar storms and other disturbances, but variations in solar activity, particularly the frequency of sunspots, are also documented at decadal to millennial timescales and probably occur at longer timescales as well. The "Maunder minimum," a period of drastically reduced sunspot activity between AD 1645 and 1715, has been suggested as a contributing factor to the Little Ice Age.

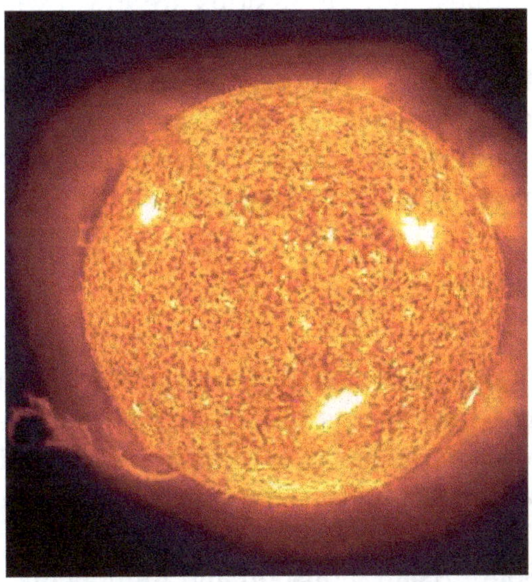

The Sun as imaged in extreme ultraviolet light by the Earth-orbiting Solar and Heliospheric Observatory (SOHO) satellite. A massive loop-shaped eruptive prominence is visible at the lower left. Nearly white areas are the hottest; deeper reds indicate cooler temperatures.

Volcanic Activity

Volcanic activity can influence climate in a number of ways at different timescales. Individual volcanic eruptions can release large quantities of sulfur dioxide and other aerosols into the stratosphere, reducing atmospheric transparency and thus the amount of solar radiation reaching Earth's surface and troposphere. A recent example is the 1991 eruption in the Philippines of Mount Pinatubo, which had measurable influences on atmospheric circulation and heat budgets. The 1815 eruption of Mount Tambora on the island of Sumbawa had more dramatic consequences, as the spring and summer of the following year (1816, known as "the year without a summer") were unusually cold over much of the world. New England and Europe experienced snowfalls and frosts throughout the summer of 1816.

A column of gas and ash rising from Mount Pinatubo in the Philippines on June 12, 1991, just days before the volcano's climactic explosion on June 15.

Volcanoes and related phenomena, such as ocean rifting and subduction, release carbon dioxide into both the oceans and the atmosphere. Emissions are low; even a massive volcanic eruption such as Mount Pinatubo releases only a fraction of the carbon dioxide emitted by fossil-fuel combustion in a year. At geologic timescales, however, release of this greenhouse gas can have important effects. Variations in carbon dioxide release by volcanoes and ocean rifts over millions of years can alter the chemistry of the atmosphere. Such changeability in carbon dioxide concentrations probably accounts for much of the climatic variation that has taken place during the Phanerozoic Eon.

Tectonic Activity

Tectonic movements of Earth's crust have had profound effects on climate at timescales of millions to tens of millions of years. These movements have changed the shape, size, position, and elevation of the continental masses as well as the bathymetry of the oceans. Topographic and bathymetric changes in turn have had strong effects on the circulation of both the atmosphere and the oceans. For example, the uplift of the Tibetan Plateau during the Cenozoic Era affected atmospheric circulation patterns, creating the South Asian monsoon and influencing climate over much of the rest of Asia and neighbouring regions.

Tectonic activity also influences atmospheric chemistry, particularly carbon dioxide concentrations. Carbon dioxide is emitted from volcanoes and vents in rift zones and subduction zones. Variations in the rate of spreading in rift zones and the degree of volcanic activity near plate margins have influenced atmospheric carbon dioxide concentrations throughout Earth's history. Even the chemical weathering of rock constitutes an important sink for carbon dioxide. (A carbon sink is any process that removes carbon dioxide from the atmosphere by the chemical conversion of CO2 to organic or inorganic carbon compounds.) Carbonic acid, formed from carbon dioxide and water, is a reactant in dissolution of silicates and other minerals. Weathering rates are related to the mass, elevation, and exposure of bedrock. Tectonic uplift can increase all these factors and thus lead to increased weathering and carbon dioxide absorption. For example, the chemical weathering of the rising Tibetan Plateau may have played an important role in depleting the atmosphere of carbon dioxide during a global cooling period in the late Cenozoic Era.

Orbital (Milankovich) Variations

The orbital geometry of Earth is affected in predictable ways by the gravitational influences of other planets in the solar system. Three primary features of Earth's orbit are affected, each in a cyclic, or regularly recurring, manner. First, the shape of Earth's orbit around the Sun, varies from nearly circular to elliptical (eccentric), with periodicities of 100,000 and 413,000 years. Second, the tilt of Earth's axis with respect to the Sun, which is primarily responsible for Earth's seasonal climates, varies between 22.1° and 24.5° from the plane of Earth's rotation around the Sun. This variation occurs on a cycle of 41,000 years. In general, the greater the tilt, the greater the solar radiation received by hemispheres in summer and the less received in winter. The third cyclic change to Earth's orbital geometry results from two combined phenomena: (1) Earth's axis of rotation wobbles, changing the direction of the axis with respect to the Sun, and (2) the orientation of Earth's orbital ellipse rotates slowly. These two processes create a 26,000-year cycle, called precession of the equinoxes, in which the position of Earth at the equinoxes and solstices changes. Today Earth is closest to the Sun (perihelion) near the December solstice, whereas 9,000 years ago perihelion occurred near the June solstice.

These orbital variations cause changes in the latitudinal and seasonal distribution of solar radiation, which in turn drive a number of climate variations. Orbital variations play major roles in pacing glacial-interglacial and monsoonal patterns. Their influences have been identified in climatic changes over much of the Phanerozoic. For example, cyclothems—which are interbedded marine, fluvial, and coal beds characteristic of the Pennsylvanian Subperiod (318.1 million to 299 million years ago)—appear to represent Milankovitch-driven changes in mean sea level.

Greenhouse Gases

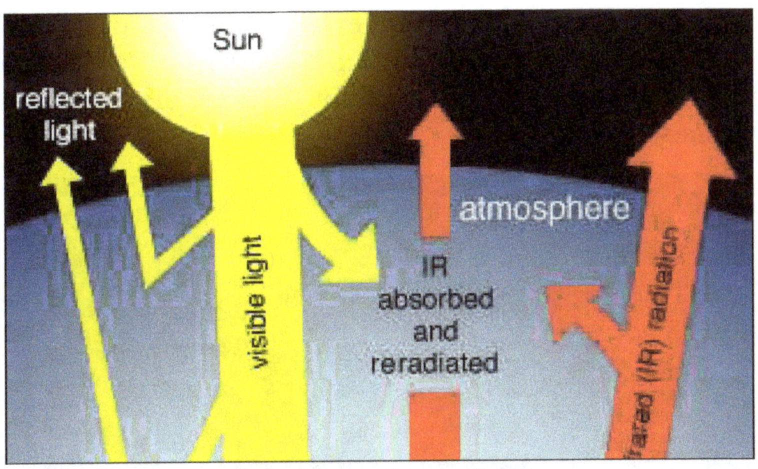

Greenhouse effect on Earth.

Greenhouse gases are gas molecules that have the property of absorbing infrared radiation (net heat energy) emitted from Earth's surface and reradiating it back to Earth's surface, thus contributing to the phenomenon known as the greenhouse effect. Carbon dioxide, methane, and water vapour are the most important greenhouse gases, and they have a profound effect on the energy budget of the Earth system despite making up only a fraction of all atmospheric gases. Concentrations of greenhouse gases have varied substantially during Earth's history, and these variations have driven substantial climate changes at a wide range of timescales.

In general, greenhouse gas concentrations have been particularly high during warm periods and low during cold phases. A number of processes influence greenhouse gas concentrations. Some, such as tectonic activities, operate at timescales of millions of years, whereas others, such as vegetation, soil, wetland, and ocean sources and sinks, operate at timescales of hundreds to thousands of years. Human activities—especially fossil-fuel combustion since the Industrial Revolution—are responsible for steady increases in atmospheric concentrations of various greenhouse gases, especially carbon dioxide, methane, ozone, and chlorofluorocarbons (CFCs).

Feedback

Perhaps the most intensively discussed and researched topic in climate variability is the role of interactions and feedbacks among the various components of the Earth system. The feedbacks involve different components that operate at different rates and timescales. Ice sheets, sea ice, terrestrial vegetation, ocean temperatures, weathering rates, ocean circulation, and greenhouse gas concentrations are all influenced either directly or indirectly by the atmosphere; however, they also all feed back into the atmosphere, thereby influencing it in important ways. For example, different forms and densities of vegetation on the land surface influence the albedo, or reflectivity, of Earth's surface, thus affecting the overall radiation budget at local to regional scales. At the same time, the transfer of water molecules from soil to the atmosphere is mediated by vegetation, both directly (from transpiration through plant stomata) and indirectly (from shading and temperature influences on direct evaporation from soil). This regulation of latent heat flux by vegetation can influence climate at local to global scales. As a result, changes in vegetation, which are partially controlled by climate, can in turn influence the climate system. Vegetation also influences greenhouse gas concentrations; living plants constitute an important sink for atmospheric carbon dioxide, whereas they act as sources of carbon dioxide when they are burned by wildfires or undergo decomposition. These and other feedbacks among the various components of the Earth system are critical for both understanding past climate changes and predicting future ones.

Mixed evergreen and hardwood forest on the slopes of the Adirondack Mountains near Keene Valley, New York.

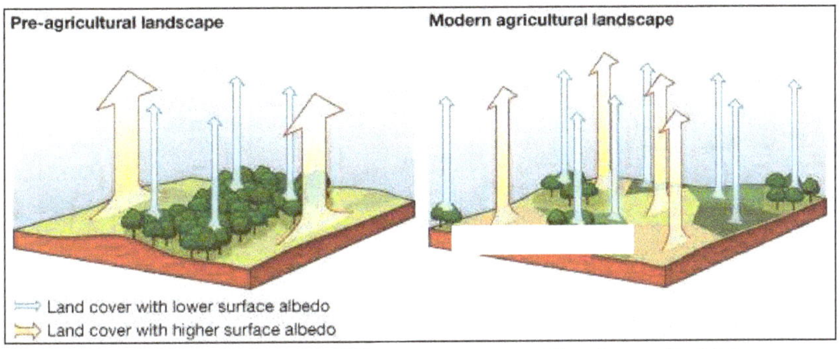

Surface reflectance (albedo) of solar energy under different patterns of land use. (Left) In a preagricultural landscape, large forest-covered areas of low surface albedo alternate with large open areas of high albedo. (Right) In an agricultural landscape, a patchwork of smaller forested and open areas exists, each with its characteristic albedo.

Human Activities

Recognition of global climate change as an environmental issue has drawn attention to the climatic impact of human activities. Most of this attention has focused on carbon dioxide emission via fossil-fuel combustion and deforestation. Human activities also yield releases of other greenhouse gases, such as methane (from rice cultivation, livestock, landfills, and other sources) and chlorofluorocarbons (from industrial sources). There is little doubt among climatologists that these greenhouse gases affect the radiation budget of Earth; the nature and magnitude of the climatic response are a subject of intense research activity. Paleoclimate records from tree rings, coral, and ice cores indicate a clear warming trend spanning the entire 20th century and the first decade of the 21st century. In fact, the 20th century was the warmest of the past 10 centuries, and the decade 2001–10 was the warmest decade since the beginning of modern instrumental record keeping. Many climatologists have pointed to this warming pattern as clear evidence of human-induced climate change resulting from the production of greenhouse gases.

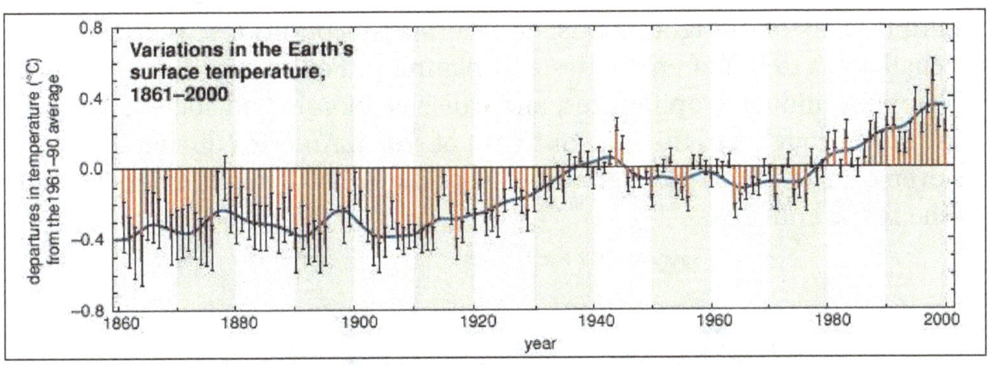

The global average surface temperature range for each year from 1861 to 2000 is shown by solid red bars, with the confidence range in the data for each year shown by thin whisker bars. The average change over time is shown by the solid curve.

A second type of human impact, the conversion of vegetation by deforestation, afforestation, and agriculture, is receiving mounting attention as a further source of climate change. It is becoming increasingly clear that human impacts on vegetation cover can have local, regional, and even global effects on climate, due to changes in the sensible and latent heat flux to the atmosphere and the distribution of energy within the climate system. The extent to which these factors contribute to recent and ongoing climate change is an important, emerging area of study.

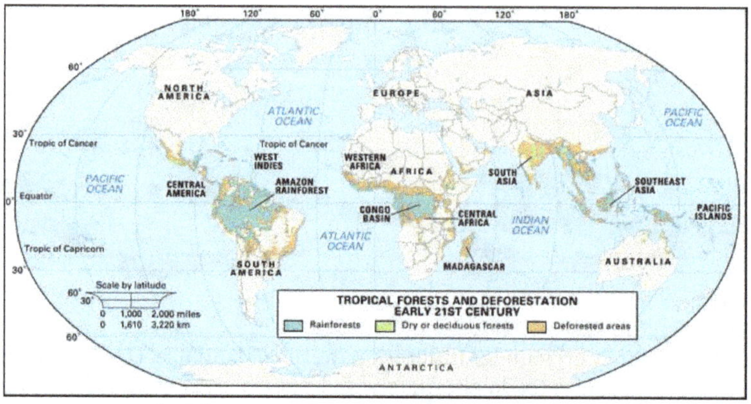

Tropical forests and deforestation. Tropical forests and deforestation in the early 21st century.

Climate Change within a Human Life Span

Regardless of their locations on the planet, all humans experience climate variability and change within their lifetimes. The most familiar and predictable phenomena are the seasonal cycles, to which people adjust their clothing, outdoor activities, thermostats, and agricultural practices. However, no two summers or winters are exactly alike in the same place; some are warmer, wetter, or stormier than others. This interannual variation in climate is partly responsible for year-to-year variations in fuel prices, crop yields, road maintenance budgets, and wildfire hazards. Single-year, precipitation-driven floods can cause severe economic damage, such as those of the upper Mississippi River drainage basin during the summer of 1993, and loss of life, such as those that devastated much of Bangladesh in the summer of 1998. Similar damage and loss of life can also occur as the result of wildfires, severe storms, hurricanes, heat waves, and other climate-related events.

Climate variation and change may also occur over longer periods, such as decades. Some locations experience multiple years of drought, floods, or other harsh conditions. Such decadal variation of climate poses challenges to human activities and planning. For example, multiyear droughts can disrupt water supplies, induce crop failures, and cause economic and social dislocation, as in the case of the Dust Bowl droughts in the midcontinent of North America during the 1930s. Multiyear droughts may even cause widespread starvation, as in the Sahel drought that occurred in northern Africa during the 1970s and '80s.

Abandoned farmstead showing the effects of wind erosion in the Dust Bowl, Texas county, Okla.

Seasonal Variation

Every place on Earth experiences seasonal variation in climate (though the shift can be slight in some tropical regions). This cyclic variation is driven by seasonal changes in the supply of solar radiation to Earth's atmosphere and surface. Earth's orbit around the Sun is elliptical; it is closer to the Sun (147 million km [about 91 million miles]) near the winter solstice and farther from the Sun (152 million km [about 94 million miles]) near the summer solstice in the Northern Hemisphere. Furthermore, Earth's axis of rotation occurs at an oblique angle (23.5°) with respect to its orbit. Thus, each hemisphere is tilted away from the Sun during its winter period and toward the Sun in its summer period. When a hemisphere is tilted away from the Sun, it receives less solar radiation than the opposite hemisphere, which at that time is pointed toward the Sun. Thus, despite the closer proximity of the Sun at the winter solstice, the Northern Hemisphere receives less solar radiation during the winter than it does during the summer. Also as a consequence of the tilt, when the Northern Hemisphere experiences winter, the Southern Hemisphere experiences summer.

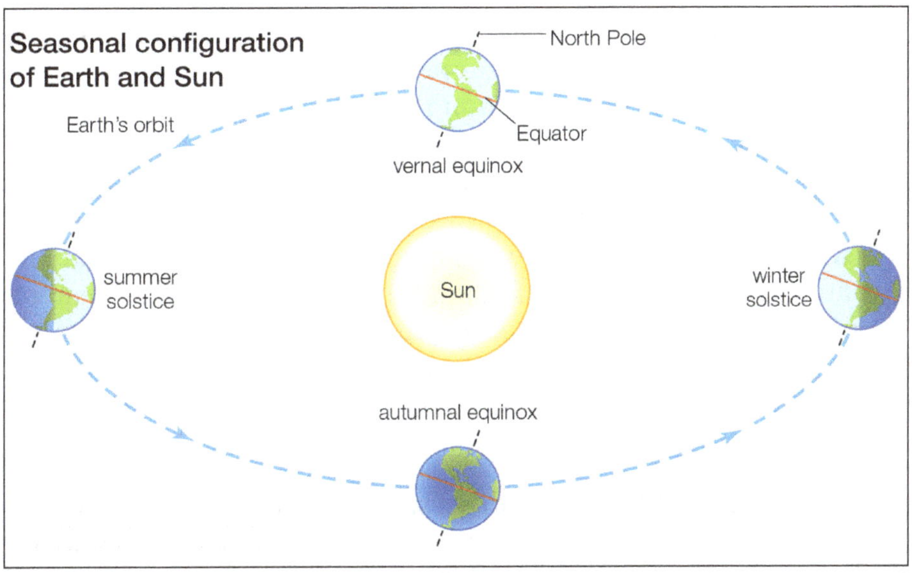

A diagram shows the position of Earth at the beginning of each season in the Northern Hemisphere.

Earth's climate system is driven by solar radiation; seasonal differences in climate ultimately result from the seasonal changes in Earth's orbit. The circulation of air in the atmosphere and water in the oceans responds to seasonal variations of available energy from the Sun. Specific seasonal changes in climate occurring at any given location on Earth's surface largely result from the transfer of energy from atmospheric and oceanic circulation. Differences in surface heating taking place between summer and winter cause storm tracks and pressure centres to shift position and strength. These heating differences also drive seasonal changes in cloudiness, precipitation, and wind.

Seasonal responses of the biosphere (especially vegetation) and cryosphere (glaciers, sea ice, snowfields) also feed into atmospheric circulation and climate. Leaf fall by deciduous trees as they go into winter dormancy increases the albedo (reflectivity) of Earth's surface and may lead to greater local and regional cooling. Similarly, snow accumulation also increases the albedo of land surfaces and often amplifies winter's effects.

Interannual Variation

Interannual climate variations, including droughts, floods, and other events, are caused by a complex array of factors and Earth system interactions. One important feature that plays a role in these variations is the periodic change of atmospheric and oceanic circulation patterns in the tropical Pacific region, collectively known as El Niño–Southern Oscillation (ENSO) variation. Although its primary climatic effects are concentrated in the tropical Pacific, ENSO has cascading effects that often extend to the Atlantic Ocean region, the interior of Europe and Asia, and the polar regions. These effects, called teleconnections, occur because alterations in low-latitude atmospheric circulation patterns in the Pacific region influence atmospheric circulation in adjacent and downstream systems. As a result, storm tracks are diverted and atmospheric pressure ridges (areas of high pressure) and troughs (areas of low pressure) are displaced from their usual patterns.

As an example, El Niño events occur when the easterly trade winds in the tropical Pacific weaken or reverse direction. This shuts down the upwelling of deep, cold waters off the west coast of South America, warms the eastern Pacific, and reverses the atmospheric pressure gradient in the western Pacific. As a result, air at the surface moves eastward from Australia and Indonesia toward the central Pacific and the Americas. These changes produce high rainfall and flash floods along the normally arid coast of Peru and severe drought in the normally wet regions of northern Australia and Indonesia. Particularly severe El Niño events lead to monsoon failure in the Indian Ocean region, resulting in intense drought in India and East Africa. At the same time, the westerlies and storm tracks are displaced toward the Equator, providing California and the desert Southwest of the United States with wet, stormy winter weather and causing winter conditions in the Pacific Northwest, which are typically wet, to become warmer and drier. Displacement of the westerlies also results in drought in northern China and from northeastern Brazil through sections of Venezuela. Long-term records of ENSO variation from historical documents, tree rings, and reef corals indicate that El Niño events occur, on average, every two to seven years. However, the frequency and intensity of these events vary through time.

The North Atlantic Oscillation (NAO) is another example of an interannual oscillation that produces important climatic effects within the Earth system and can influence climate throughout the Northern Hemisphere. This phenomenon results from variation in the pressure gradient, or the difference in atmospheric pressure between the subtropical high, usually situated between the Azores and Gibraltar, and the Icelandic low, centred between Iceland and Greenland. When the pressure gradient is steep due to a strong subtropical high and a deep Icelandic low (positive phase), northern Europe and northern Asia experience warm, wet winters with frequent strong winter storms. At the same time, southern Europe is dry. The eastern United States also experiences warmer, less snowy winters during positive NAO phases, although the effect is not as great as in Europe. The pressure gradient is dampened when NAO is in a negative mode—that is, when a weaker pressure gradient exists from the presence of a weak subtropical high and Icelandic low. When this happens, the Mediterranean region receives abundant winter rainfall, while northern Europe is cold and dry. The eastern United States is typically colder and snowier during a negative NAO phase.

During years when the North Atlantic Oscillation (NAO) is in its positive phase, the eastern United States, southeastern Canada, and northwestern Europe experience warmer winter temperatures, whereas colder temperatures are found in these locations during its negative phase. When the El Niño/Southern Oscillation (ENSO) and NAO are both in their positive phase, European winters tend to be wetter and less severe; however, beyond this general tendency, the influence of the ENSO upon the NAO is not well understood.

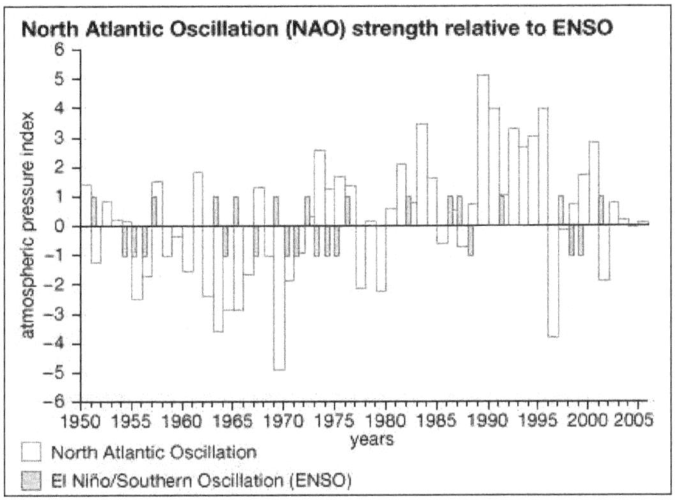

The ENSO and NAO cycles are driven by feedbacks and interactions between the oceans and atmosphere. Interannual climate variation is driven by these and other cycles, interactions among cycles, and perturbations in the Earth system, such as those resulting from large injections of aerosols from volcanic eruptions. One example of a perturbation due to volcanism is the 1991 eruption of Mount Pinatubo in the Philippines, which led to a decrease in the average global temperature of approximately 0.5 °C (0.9 °F) the following summer.

Decadal Variation

Climate varies on decadal timescales, with multiyear clusters of wet, dry, cool, or warm conditions. These multiyear clusters can have dramatic effects on human activities and welfare. For instance, a severe three-year drought in the late 16th century probably contributed to the destruction of Sir Walter Raleigh's "Lost Colony" at Roanoke Island in what is now North Carolina, and a subsequent seven-year drought (1606–12) led to high mortality at the Jamestown Colony in Virginia. Also, some scholars have implicated persistent and severe droughts as the main reason for the collapse of the Maya civilization in Mesoamerica between AD 750 and 950; however, discoveries in the early 21st century suggest that war-related trade disruptions played a role, possibly interacting with famines and other drought-related stresses.

Although decadal-scale climate variation is well documented, the causes are not entirely clear. Much decadal variation in climate is related to interannual variations. For example, the frequency and magnitude of ENSO change through time. The early 1990s were characterized by repeated El Niño events, and several such clusters have been identified as having taken place during the 20th century. The steepness of the NAO gradient also changes at decadal timescales; it has been particularly steep since the 1970s.

Recent research has revealed that decadal-scale variations in climate result from interactions between the ocean and the atmosphere. One such variation is the Pacific Decadal Oscillation (PDO), also referred to as the Pacific Decadal Variability (PDV), which involves changing sea surface temperatures (SSTs) in the North Pacific Ocean. The SSTs influence the strength and position of the

Aleutian Low, which in turn strongly affects precipitation patterns along the Pacific Coast of North America. PDO variation consists of an alternation between "cool-phase" periods, when coastal Alaska is relatively dry and the Pacific Northwest relatively wet (e.g., 1947–76), and "warm-phase" periods, characterized by relatively high precipitation in coastal Alaska and low precipitation in the Pacific Northwest (e.g., 1925–46, 1977–98). Tree ring and coral records, which span at least the last four centuries, document PDO variation.

A similar oscillation, the Atlantic Multidecadal Oscillation (AMO), occurs in the North Atlantic and strongly influences precipitation patterns in eastern and central North America. A warm-phase AMO (relatively warm North Atlantic SSTs) is associated with relatively high rainfall in Florida and low rainfall in much of the Ohio Valley. However, the AMO interacts with the PDO, and both interact with interannual variations, such as ENSO and NAO, in complex ways . Such interactions may lead to the amplification of droughts, floods, or other climatic anomalies. For example, severe droughts over much of the conterminous United States in the first few years of the 21st century were associated with warm-phase AMO combined with cool-phase PDO. The mechanisms underlying decadal variations, such as PDO and AMO, are poorly understood, but they are probably related to ocean-atmosphere interactions with larger time constants than interannual variations. Decadal climatic variations are the subject of intense study by climatologists and paleoclimatologists.

OZONE DEPLETION AND CLIMATE CHANGE

Ozone depletion and climate change, or Ozone hole and global warming in more popular terms, are environmental challenges whose connections have been explored and which have been compared and contrasted, for example in terms of global regulation, in various studies and books.

There is widespread scientific interest in better regulation of climate change, ozone depletion and air pollution, as in general the human relationship with the biosphere is deemed of major historiographical and political significance. Already by 1994 the legal debates about respective regulation regimes on climate change, ozone depletion and air pollution were being dubbed "monumental" and a combined synopsis provided.

There are some parallels between atmospheric chemistry and anthropogenic emissions in the discussions which have taken place and the regulatory attempts which have been made. Most important is that the gases causing both problems have long lifetimes after emission to the atmosphere, thus causing problems which are difficult to reverse. However, the Vienna Convention for the Protection of the Ozone Layer and the Montreal Protocol that amended it are seen as success stories, while the Kyoto Protocol on anthropogenic climate change has largely failed. Currently efforts are being undertaken to assess the reasons and to use synergies, for example with regard to data reporting and policy design and further exchanging of information. While the general public tends to see global warming as a subset of ozone depletion, in fact ozone and chemicals such as chlorofluorocarbons (CFCs) and other halocarbons, which are held responsible for ozone depletion, are important greenhouse gasses. Furthermore, natural levels of ozone in both the stratosphere and troposphere have a warming effect.

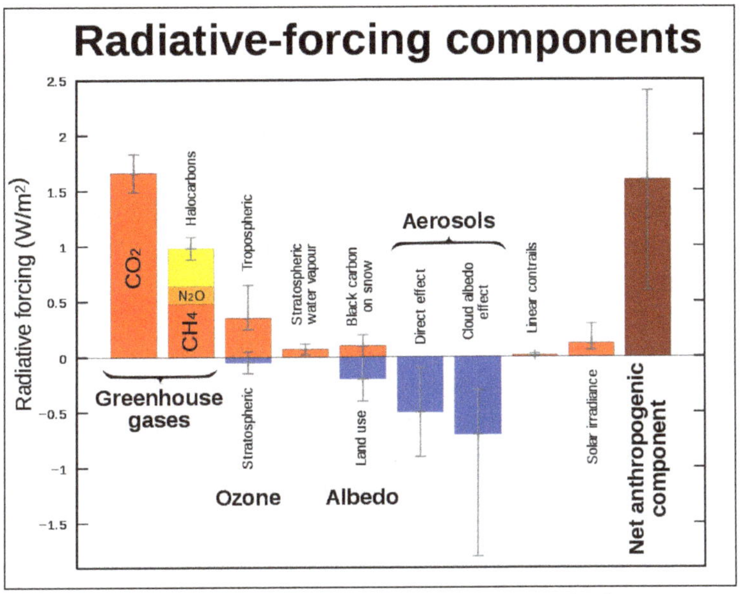

Radiative forcing from various greenhouse gases and other sources.

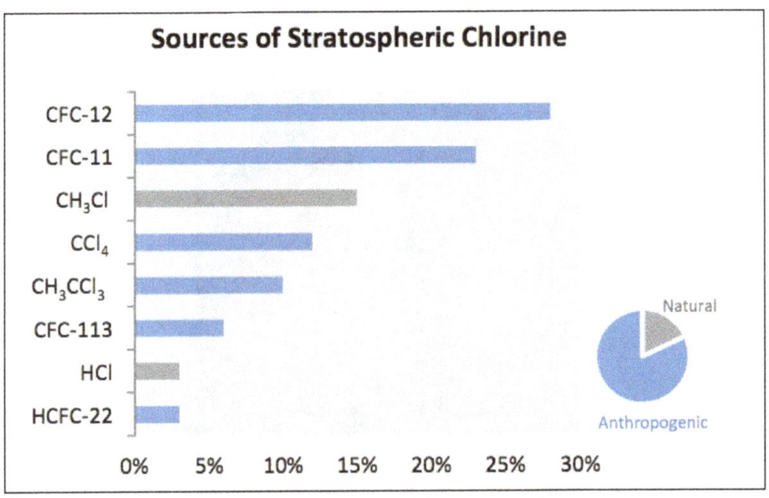

Sources of Stratospheric Chlorine.

There are various links between the two fields of human-atmospheric interaction. Policy experts have advocated for a closer linking of ozone protection and climate protection efforts.

Drew Shindell has used climate models to assess both climate change and ozone depletion. In his view, while research up to now has been more about the impact of CFC emissions on stratospheric ozone, the future will be more about the interaction between climate change and ozone feedback. Ozone is a greenhouse gas itself. Many ozone-depleting substances are also greenhouse gases, some agents of radiative forcing are thousands of times more powerful than carbon dioxide over the short and medium term. The increases in concentrations of these chemicals have produced 0.34 ± 0.03 W/m^2 of radiative forcing, corresponding to about 14% of the total radiative forcing from increases in the concentrations of well-mixed greenhouse gases. Already the natural ozone variability in the stratosphere seems to be closely correlated with the 11-year solar cycle of irradiance changes and has, via a dynamic coupling between the stratosphere and troposphere, a significant impact on climate.

As with carbon dioxide and methane, there are some natural sources of tropospheric chlorine, such as sea spray. Chlorine from ocean spray is soluble and thus is washed by rainfall before it reaches the stratosphere. It is *stratospheric* chlorine that affects ozone depletion. Only methyl chloride, which is one of the halocarbons, has a mainly natural source, and it is responsible for about 20% of the chlorine in the stratosphere; the remaining 80% comes from man-made sources. Chlorofluorocarbons, in contrast, are insoluble and long-lived, allowing them to reach the stratosphere. In the lower atmosphere, there is much more chlorine from CFCs and related haloalkanes than there is in hydrogen chloride from salt spray, and in the stratosphere halocarbons are dominant.

The same CO_2 radiative forcing that produces global warming is expected to cool the stratosphere. This cooling, in turn, is expected to produce a relative *increase* in ozone (O_3) depletion in the polar area and in the frequency of ozone holes. Conversely, ozone depletion represents a radiative forcing of the climate system of about -0.15 ± 0.10 watts per square metre (W/m^2).

Policy Approach

Sir Robert (Bob) Watson played an important role in both cases.

There are both links and major differences between ozone depletion and global warming and the way the two challenges have been handled. While in the case of atmospheric ozone depletion, in a situation of high uncertainty and against strong resistance, climate change regulation attempts at the international level such as the Kyoto Protocol have failed to reduce global emissions. The Vienna Convention for the Protection of the Ozone Layer and the Montreal Protocol were both originally signed by only some member states of the United Nations (43 nations in the case of the Montreal Protocol in 1986) while Kyoto attempted to create a worldwide agreement from scratch. Expert consensus concerning CFCs in the form of the Scientific Assessment of Ozone Depletion was reached long after the first regulatory steps were taken, and as of 29 December 2012, all countries in the United Nations plus the Cook Islands, the Holy See, Niue and the supranational European Union had ratified the original Montreal Protocol. These countries have also ratified the London, Copenhagen, and Montreal amendments to the Protocol. As of 15 April 2014, the Beijing amendments had not been ratified by two state parties.

After the Vienna Convention, the halocarbon industry shifted its position and started supporting a protocol to limit CFC production. US manufacturer DuPont acted more quickly than their European counterparts. The EU shifted its position as well after Germany, which has a substantial chemical industry, gave up its defence of the CFC industry and started supporting more regulation. Government and industry in France and the UK had tried to defend their CFC-producing industries even after the Montreal Protocol had been signed.

The Vienna Convention was installed before a scientific consensus on the ozone hole was established. On the contrary, until the 1980s the EU, NASA, NAS, UNEP, WMO and the British government had issued scientific reports with divergent conclusions. Sir Robert (Bob) Watson, Director of the Science Division at NASA, played a crucial role in the process of reaching a unified assessment.

Policy and Consensus

Bob Watson successfully united the international science community in 1985 to act on the problem with the ozone hole, before a consensus existed.

Aant Elzinga wrote in 1996 about the consensus, that the Intergovernmental Panel on Climate Change has tried in the prior two reports a global consensus approach to climate action. Stephen Schneider and Paul N. Edwards, noted in 1997, that after the IPCC Second Assessment Report, the lobby group Global Climate Coalition and a few self-proclaimed "contrarian" scientists tried to discredit the conclusions of the report. They pointed out that the goal of the IPCC is to fairly represent the complete range of credible scientific opinion and if possible a consensus view.

In 2007, Reiner Grundmann compared climate actions in Europe and the United States, he interpreted the inaction besides existing consensus, and noted, *Political agenda that drove US climate change policy. The high visibility of sceptical scientists in the media resonates with this*, and wrote that Germany started ambitious goals, reduced emissions, because 'balanced reporting' led to a bias in climate change coverage in advantage of sceptical arguments in the U.S., but not so much in Germany. Additionally, Grundmann pointed out that after warnings from scientists in 1986 the German Parliament commissioned the *Enquetekommission 'Vorsorge zum Schutz der Erdatmosphäre'* (Precaution for the Protection of the Earth's Atmosphere), to assess the situation, consisting of scientists, politicians and representatives of interest groups. Three years later the report made an impact with the assessment of the state of the art in climate research, an assessment of the threat of climate change itself as well as suggestions for clear emissions reduction targets, even though he argues there was no consensus, and attributed the success of the report to strong precautionary action, and that no scientific outsiders or climate change skeptics were involved.

A linear model of policy-making, based on a position that "the more knowledge we have, the better the political response will be", was not applied in the ozone case. On the contrary, the CFC regulation process focused more on managing ignorance and uncertainties as a basis of political decision making, as the relationships between science, public (lack of) understanding and policy were better taken into account. In the meantime, such a player in the IPCC process as Michael Oppenheimer conceded some limitations of the IPCC consensus approach and asked for concurring, smaller assessments of special problems instead of repetitions of the large-scale approach every six years. It has become more important to provide a broader exploration of uncertainties. Others also see mixed blessings in the drive for consensus within the IPCC process and have asked for dissenting or minority positions to be included or for statements about uncertainties to be improved.

Public Opinion

The two atmospheric problems have achieved significantly different levels of understanding by the public, including both the basic science and policy issues. People have limited scientific knowledge about global warming and tend to confuse it with or see it as a subset of the ozone hole. Not only on the policy level, ozone regulation fared much better than climate change in public opinion. Americans voluntarily switched away from aerosol sprays before legislation was enforced, while climate change has failed in achieving a broader scientific comprehension and in raising comparable concern.

The metaphors used in the CFC discussion (ozone shield, ozone hole) resonated better with non-scientists and their concerns. The ozone case was communicated to lay persons «with easy-to-understand bridging metaphors derived from the popular culture" and related to "immediate risks with everyday relevance", while the public opinion on climate change sees no imminent danger. The ozone hole was much more seen as a «hot issue» and imminent risk compared to global climate change, as lay people feared a depletion of the ozone layer (ozone shield) risked increasing severe consequences such skin cancer, cataracts, damage to plants, and reduction of plankton populations in the ocean›s photic zone. This was not the case with global warming.

Personal Risk Assessment and Knowledge

Sheldon Ungar, a Canadian sociologist, assumes that while the quantity of specialized knowledge is exploding, in contrast scientific ignorance among lay people is the norm and even increasing. Public opinion failed to tie climate change to concrete events which could be used as a threshold or beacon to signify immediate danger. Scientific predictions of a temperature rise of 2 °C (4 °F) to 3 °C (5 °F) over several decades do not resonate with people, for example in North America, who experience similar swings during a single day. As scientists define global warming as a problem of the future, a liability in the «attention economy», pessimistic outlooks in general and the attribution of extreme weather to climate change have often been discredited or ridiculed in the public arena (compare the Gore effect). Even when James Hansen tried to use the 1988–89 North American drought as a call to action, scientists kept stating, in line with the IPCC findings, that even extreme weather is not climate. While the greenhouse effect, *per se*, is essential for life on earth, the case was quite different with the ozone hole and other metaphors about ozone depletion. The scientific assessment of the ozone problem also had large uncertainties; both the ozone content of the upper atmosphere and its depletion are complicated to measure and the link between ozone depletion and rates of enhanced skin cancer is rather weak. But the metaphors used in the discussion (ozone shield, ozone hole) resonated better with lay people and their concerns.

Layers of the atmosphere (not to scale). The Earth's ozone layer is mainly found in the lower portion of the stratosphere from approximately 20 to 30 kilometres (12 to 19 mi) above Earth.

> The idea of rays penetrating a damaged "shield" meshes nicely with abiding and resonant cultural motifs, including "Hollywood affinities." These range from the shields on the Starship Enterprise to Star Wars. It is these pre-scientific bridging metaphors built around the penetration of a deteriorating shield that render the ozone problem relatively simple. That the ozone threat can be linked with Darth Vader means that it is encompassed in common sense understandings that are deeply ingrained and widely shared.
>
> —*Sheldon Ungar*

The CFC regulation attempts at the end of the 1980s profited from those easy to grasp metaphors and the personal risk assumptions taken from them. The fate of celebrities like President Ronald Reagan, who had skin cancer removal from his nose in 1985 and 1987, was also of high importance. In case of the public opinion on climate change, no imminent danger is perceived.

Cost-Benefit Assessments and Industry Policy

Cass Sunstein and others have compared the differing approach of the United States to the Montreal Protocol, which it accepted, and the Kyoto Protocol, which it rejected. Sunstein assumes that the cost-benefit assessments of climate change action for the US were instrumental in the US› withdrawal from participation in Kyoto. Daniel Magraw, also a lawyer, considers governmental motivations besides relative costs and benefits as being of higher importance. Peter Orszag and Terry Dinan took an insurance perspective and assume that an assessment which predicted dire consequences of climate change would be more of a motivation for the US to change its stance on global warming and adopting regulation measurements.

The US chemical company DuPont had already lost some of their zeal in defending their products after a strategic manufacturing patent for Freon was set to expire in 1979. A citizen boycott of spray cans gained importance in parallel. Not by chance, the United States banned the use of CFCs in aerosol cans in 1978.

Government and industry in France and the UK tried to defend their CFC-producing industries even after the Montreal Protocol had been signed. The European Community rejected proposals to ban CFCs in aerosol sprays for a long time. The EU shifted its position after Germany, which also has a large chemical industry, gave up its defence of the CFC industry and started supporting moves towards regulation. After regulation was more and more enforced, DuPont acted faster than their European counterparts as they may have feared court action related to increased skin cancer, especially as the EPA had published a study in 1986 claiming that an additional 40 million cases and 800,000 cancer deaths were to be expected in the US in the next 88 years. The identification and marketing of a 100% ozone-safe hydrocarbon refrigerant called «Greenfreeze" by the NGO Greenpeace in the early 1990s had a rapid significant impact in major markets of Europe and Asia. The climate change protocols were less successful. In the case of Kyoto, then secretary of the environment Angela Merkel, prevented a possible failure by suggesting to use 1990 as starting date for emission reduction. In so far the demise of the Eastern European heavy industry allowed for a high commitment, but actual emissions kept on growing on a global scale.

GLOBAL WARMING

Global warming is a phenomenon of increasing average air temperatures near the surface of Earth over the past one to two centuries. Climate scientists have since the mid-20th century gathered detailed observations of various weather phenomena (such as temperatures, precipitation, and storms) and of related influences on climate (such as ocean currents and the atmosphere's chemical composition). These data indicate that Earth's climate has changed over almost every conceivable timescale since the beginning of geologic time and that the influence of human activities since at least the beginning of the Industrial Revolution has been deeply woven into the very fabric of climate change.

Giving voice to a growing conviction of most of the scientific community, the Intergovernmental Panel on Climate Change (IPCC) was formed in 1988 by the World Meteorological Organization (WMO) and the United Nations Environment Program (UNEP). In 2013 the IPCC reported that the interval between 1880 and 2012 saw an increase in global average surface temperature of approximately 0.9 °C (1.5 °F). The increase is closer to 1.1 °C (2.0 °F) when measured relative to the preindustrial (i.e., 1750–1800) mean temperature.

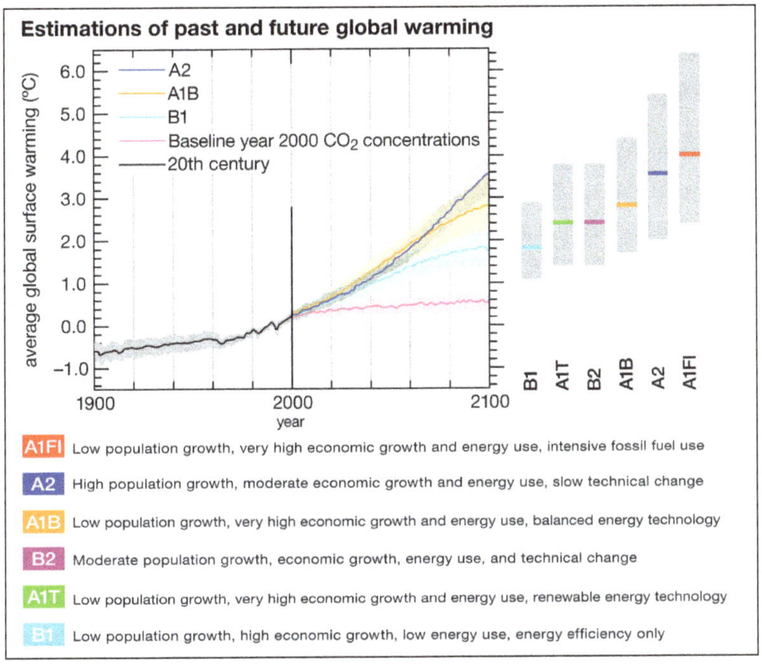

Global warming scenarios.

Graph of the predicted increase in Earth's average surface temperature according to a series of climate change scenarios that assume different levels of economic development, population growth, and fossil fuel use. The assumptions made by each scenario are given at the bottom of the graph.

A special report produced by the IPCC in 2018 honed this estimate further, noting that human beings and human activities have been responsible for a worldwide average temperature increase of between 0.8 and 1.2 °C (1.4 and 2.2 °F) of global warming since preindustrial times, and most of the warming observed over the second half of the 20th century could be attributed to human activities. It predicted that the global mean surface temperature would increase between 3 and 4 °C (5.4 and 7.2 °F) by 2100 relative to the 1986–2005 average should carbon emissions continue at their current rate. The predicted rise in temperature was based on a range of possible scenarios that accounted for future greenhouse gas emissions and mitigation (severity reduction) measures and on uncertainties in the model projections. Some of the main uncertainties include the precise role of feedback processes and the impacts of industrial pollutants known as aerosols, which may offset some warming.

Many climate scientists agree that significant societal, economic, and ecological damage would result if global average temperatures rose by more than 2 °C (3.6 °F) in such a short time. Such damage would include increased extinction of many plant and animal species, shifts in patterns of agriculture, and rising sea levels. By 2015 all but a few national governments had begun the process of instituting carbon reduction plans as part of the Paris Agreement, a treaty designed to help countries keep global warming to 1.5 °C (2.7 °F) above preindustrial levels in order to avoid the worst of the predicted effects. Authors of a special report published by the IPCC in 2018 noted that should carbon emissions continue at their present rate, the increase in average near-surface air temperatures would reach 1.5 °C sometime between 2030 and 2052. Past IPCC assessments reported that the global average sea level rose by some 19–21 cm (7.5–8.3 inches) between 1901 and 2010 and that sea levels rose faster in the second half of the 20th century than in the first half. It also predicted, again depending on a wide range of scenarios, that the global average sea level would rise 26–77 cm (10.2–30.3 inches) relative to the 1986–2005 average by 2100 for global warming of 1.5 °C, an average of 10 cm (3.9 inches) less than what would be expected if warming rose to 2 °C (3.6 °F) above preindustrial levels.

The scenarios referred to above depend mainly on future concentrations of certain trace gases, called greenhouse gases, that have been injected into the lower atmosphere in increasing amounts through the burning of fossil fuels for industry, transportation, and residential uses. Modern global warming is the result of an increase in magnitude of the so-called greenhouse effect, a warming of Earth's surface and lower atmosphere caused by the presence of water vapour, carbon dioxide, methane, nitrous oxides, and other greenhouse gases. In 2014 the IPCC reported that concentrations of carbon dioxide, methane, and nitrous oxides in the atmosphere surpassed those found in ice cores dating back 800,000 years.

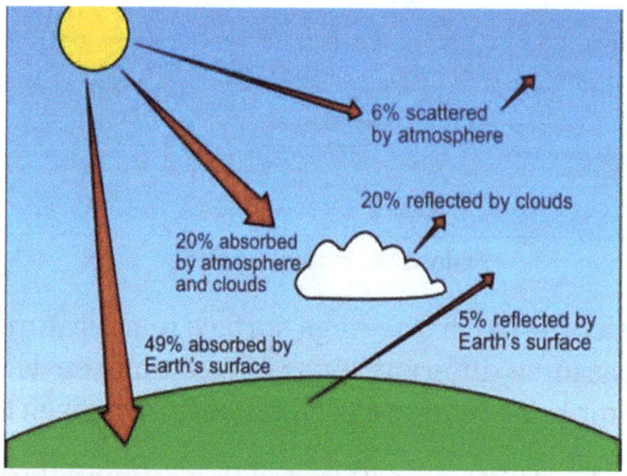

The greenhouse effect on Earth. Some incoming sunlight is reflected by Earth's atmosphere and surface, but most is absorbed by the surface, which is warmed. Infrared (IR) radiation is then emitted from the surface. Some IR radiation escapes to space, but some is absorbed by the atmosphere's greenhouse gases (especially water vapour, carbon dioxide, and methane) and reradiated in all directions, some to space and some back toward the surface, where it further warms the surface and the lower atmosphere.

Of all these gases, carbon dioxide is the most important, both for its role in the greenhouse effect and for its role in the human economy. It has been estimated that, at the beginning of the industrial age in the mid-18th century, carbon dioxide concentrations in the atmosphere were roughly 280 parts per million (ppm). By the middle of 2018 they had risen to 406 ppm, and, if fossil fuels continue to be burned at current rates, they are projected to reach 550 ppm by the mid-21st century—essentially, a doubling of carbon dioxide concentrations in 300 years.

A vigorous debate is in progress over the extent and seriousness of rising surface temperatures, the effects of past and future warming on human life, and the need for action to reduce future warming and deal with its consequences.

Climatic Variation Since the Last Glaciation

Global warming is related to the more general phenomenon of climate change, which refers to changes in the totality of attributes that define climate. In addition to changes in air temperature, climate change involves changes to precipitation patterns, winds, ocean currents, and other measures of Earth's climate. Normally, climate change can be viewed as the combination of various natural forces occurring over diverse timescales. Since the advent of human civilization, climate change has involved an "anthropogenic," or exclusively human-caused, element, and this anthropogenic element has become more important in the industrial period of the past two centuries. The term global warming is used specifically to refer to any warming of near-surface air during the past two centuries that can be traced to anthropogenic causes.

A series of photographs of the Grinnell Glacier taken from the summit of Mount Gould in Glacier National Park, Montana, in 1938, 1981, 1998, and 2006 (from left to right). In 1938 the Grinnell Glacier filled the entire area at the bottom of the image. By 2006 it had largely disappeared from this view.

To define the concepts of global warming and climate change properly, it is first necessary to recognize that the climate of Earth has varied across many timescales, ranging from an individual human life span to billions of years. This variable climate history is typically classified in terms of "regimes" or "epochs." For instance, the Pleistocene glacial epoch (about 2,600,000 to 11,700 years ago) was marked by substantial variations in the global extent of glaciers and ice sheets. These variations took place on timescales of tens to hundreds of millennia and were driven by changes in the distribution of solar radiation across Earth's surface. The distribution of solar radiation is known as the insolation pattern, and it is strongly affected by the geometry of Earth's orbit around the Sun and by the orientation, or tilt, of Earth's axis relative to the direct rays of the Sun.

Worldwide, the most recent glacial period, or ice age, culminated about 21,000 years ago in what is often called the Last Glacial Maximum. During this time, continental ice sheets extended well into the middle latitude regions of Europe and North America, reaching as far south as present-day London and New York City. Global annual mean temperature appears to have been about 4–5 °C (7–9 °F) colder than in the mid-20th century. It is important to remember that these figures are a global average. In fact, during the height of this last ice age, Earth's climate was characterized by greater cooling at higher latitudes (that is, toward the poles) and relatively little cooling over large parts of the tropical oceans (near the Equator). This glacial interval terminated abruptly about 11,700 years ago and was followed by the subsequent relatively ice-free period known as the Holocene Epoch. The modern period of Earth's history is conventionally defined as residing within the Holocene. However, some scientists have argued that the Holocene Epoch terminated in the relatively recent past and that Earth currently resides in a climatic interval that could justly be called the Anthropocene Epoch—that is, a period during which humans have exerted a dominant influence over climate.

Though less dramatic than the climate changes that occurred during the Pleistocene Epoch, significant variations in global climate have nonetheless taken place over the course of the Holocene. During the early Holocene, roughly 9,000 years ago, atmospheric circulation and precipitation patterns appear to have been substantially different from those of today. For example, there is evidence for relatively wet conditions in what is now the Sahara Desert. The change from one climatic regime to another was caused by only modest changes in the pattern of insolation within the Holocene interval as well as the interaction of these patterns with large-scale climate phenomena such as monsoons and El Niño/Southern Oscillation (ENSO).

During the middle Holocene, some 5,000–7,000 years ago, conditions appear to have been relatively warm—indeed, perhaps warmer than today in some parts of the world and during certain seasons. For this reason, this interval is sometimes referred to as the Mid-Holocene Climatic Optimum. The relative warmth of average near-surface air temperatures at this time, however, is somewhat unclear. Changes in the pattern of insolation favoured warmer summers at higher latitudes in the Northern Hemisphere, but these changes also produced cooler winters in the Northern Hemisphere and relatively cool conditions year-round in the tropics. Any overall hemispheric or global mean temperature changes thus reflected a balance between competing seasonal and regional changes. In fact, recent theoretical climate model studies suggest that global mean temperatures during the middle Holocene were probably 0.2–0.3 °C (0.4–0.5 °F) colder than average late 20th-century conditions.

Over subsequent millennia, conditions appear to have cooled relative to middle Holocene levels. This period has sometimes been referred to as the "Neoglacial." In the middle latitudes this cooling trend was associated with intermittent periods of advancing and retreating mountain glaciers reminiscent of (though far more modest than) the more substantial advance and retreat of the major continental ice sheets of the Pleistocene climate epoch.

Causes of Global Warming

The Greenhouse Effect

The average surface temperature of Earth is maintained by a balance of various forms of solar and terrestrial radiation. Solar radiation is often called "shortwave" radiation because the frequencies of the radiation are relatively high and the wavelengths relatively short—close to the visible portion of the electromagnetic spectrum. Terrestrial radiation, on the other hand, is often called "longwave" radiation because the frequencies are relatively low and the wavelengths relatively long—somewhere in the infrared part of the spectrum. Downward-moving solar energy is typically measured in watts per square metre. The energy of the total incoming solar radiation at the top of Earth's atmosphere (the so-called "solar constant") amounts roughly to 1,366 watts per square metre annually. Adjusting for the fact that only one-half of the planet's surface receives solar radiation at any given time, the average surface insolation is 342 watts per square metre annually.

The amount of solar radiation absorbed by Earth's surface is only a small fraction of the total solar radiation entering the atmosphere. For every 100 units of incoming solar radiation, roughly 30 units are reflected back to space by either clouds, the atmosphere, or reflective regions of Earth's surface. This reflective capacity is referred to as Earth's planetary albedo, and it need not remain fixed over time, since the spatial extent and distribution of reflective formations, such as clouds and ice cover, can change. The 70 units of solar radiation that are not reflected may be absorbed by the atmosphere, clouds, or the surface. In the absence of further complications, in order to maintain thermodynamic equilibrium, Earth's surface and atmosphere must radiate these same 70 units back to space. Earth's surface temperature (and that of the lower layer of the atmosphere essentially in contact with the surface) is tied to the magnitude of this emission of outgoing radiation according to the Stefan-Boltzmann law.

Earth's energy budget is further complicated by the greenhouse effect. Trace gases with certain chemical properties—the so-called greenhouse gases, mainly carbon dioxide (CO_2), methane (CH_4), and nitrous oxide (N_2O)—absorb some of the infrared radiation produced by Earth's surface. Because of this absorption, some fraction of the original 70 units does not directly escape to space. Because greenhouse gases emit the same amount of radiation they absorb and because this radiation is emitted equally in all directions (that is, as much downward as upward), the net effect of absorption by greenhouse gases is to increase the total amount of radiation emitted downward toward Earth's surface and lower atmosphere. To maintain equilibrium, Earth's surface and lower atmosphere must emit more radiation than the original 70 units. Consequently, the surface temperature must be higher. This process is not quite the same as that which governs a true greenhouse, but the end effect is similar. The presence of greenhouse gases in the atmosphere leads to a warming of the surface and lower part of the atmosphere (and a cooling higher up in the atmosphere) relative to what would be expected in the absence of greenhouse gases.

It is essential to distinguish the "natural," or background, greenhouse effect from the "enhanced" greenhouse effect associated with human activity. The natural greenhouse effect is associated with surface warming properties of natural constituents of Earth's atmosphere, especially water vapour, carbon dioxide, and methane. The existence of this effect is accepted by all scientists. Indeed, in its absence, Earth's average temperature would be approximately 33 °C (59 °F) colder than today, and Earth would be a frozen and likely uninhabitable planet. What has been subject to controversy is the so-called enhanced greenhouse effect, which is associated with increased concentrations of greenhouse gases caused by human activity. In particular, the burning of fossil fuels raises the concentrations of the major greenhouse gases in the atmosphere, and these higher concentrations have the potential to warm the atmosphere by several degrees.

Radiative Forcing

It is apparent that the temperature of Earth's surface and lower atmosphere may be modified in three ways: (1) through a net increase in the solar radiation entering at the top of Earth's atmosphere, (2) through a change in the fraction of the radiation reaching the surface, and (3) through a change in the concentration of greenhouse gases in the atmosphere. In each case the changes can be thought of in terms of "radiative forcing." As defined by the IPCC, radiative forcing is a measure of the influence a given climatic factor has on the amount of downward-directed radiant energy impinging upon Earth's surface. Climatic factors are divided between those caused primarily by human activity (such as greenhouse gas emissions and aerosol emissions) and those caused by natural forces (such as solar irradiance); then, for each factor, so-called forcing values are calculated for the time period between 1750 and the present day. "Positive forcing" is exerted by climatic factors that contribute to the warming of Earth's surface, whereas "negative forcing" is exerted by factors that cool Earth's surface.

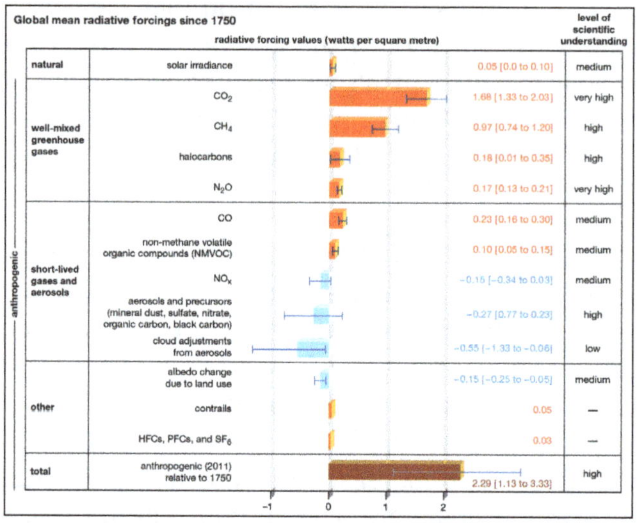

Since 1750 the concentration of carbon dioxide and other greenhouse gases has increased in Earth's atmosphere. As a result of these and other factors, Earth's atmosphere retains more heat than in the past.

On average, about 342 watts of solar radiation strike each square metre of Earth's surface per year, and this quantity can in turn be related to a rise or fall in Earth's surface temperature. Temperatures at the surface may also rise or fall through a change in the distribution of terrestrial radiation (that is, radiation emitted by Earth) within the atmosphere. In some cases, radiative forcing has a natural origin, such as during explosive eruptions from volcanoes where vented gases and ash block some portion of solar radiation from the surface. In other cases, radiative forcing has an anthropogenic, or exclusively human, origin. For example, anthropogenic increases in carbon dioxide, methane, and nitrous oxide are estimated to account for 2.3 watts per square metre of positive radiative forcing. When all values of positive and negative radiative forcing are taken together and all interactions between climatic factors are accounted for, the total net increase in surface radiation due to human activities since the beginning of the Industrial Revolution is 1.6 watts per square metre.

The Influences of Human Activity on Climate

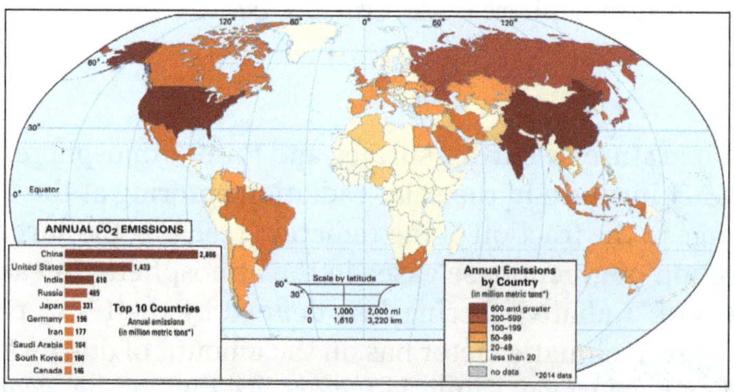

Carbon dioxide emissions: Map of annual carbon dioxide emissions by country.

Human activity has influenced global surface temperatures by changing the radiative balance governing the Earth on various timescales and at varying spatial scales. The most profound and well-known anthropogenic influence is the elevation of concentrations of greenhouse gases in the atmosphere. Humans also influence climate by changing the concentrations of aerosols and ozone and by modifying the land cover of Earth's surface.

Petroleum refinery at Ras Tanura, Saudi Arabia.Herbert Lanks/Shostal Associates.

Natural gas facility near Kursk, Russia.

Greenhouse Gases

Greenhouse gases warm Earth's surface by increasing the net downward longwave radiation reaching the surface. The relationship between atmospheric concentration of greenhouse gases and the associated positive radiative forcing of the surface is different for each gas. A complicated relationship exists between the chemical properties of each greenhouse gas and the relative amount of longwave radiation that each can absorb.

Factories that burn fossil fuels help to cause global warming.

Water Vapour

Water vapour is the most potent of the greenhouse gases in Earth's atmosphere, but its behaviour is fundamentally different from that of the other greenhouse gases. The primary role of water vapour is not as a direct agent of radiative forcing but rather as a climate feedback—that is, as a response within the climate system that influences the system's continued activity. This distinction arises from the fact that the amount of water vapour in the atmosphere cannot, in general, be directly modified by human behaviour but is instead set by air temperatures. The warmer the surface, the greater the evaporation rate of water from the surface. As a result, increased evaporation leads to a greater concentration of water vapour in the lower atmosphere capable of absorbing longwave radiation and emitting it downward.

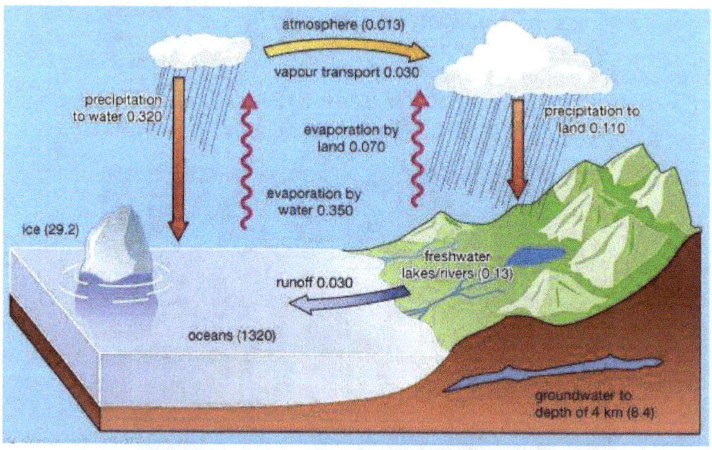

The present-day surface hydrologic cycle, in which water is transferred from the oceans through the atmosphere to the continents and back to the oceans over and beneath the land surface. The values in parentheses following the various forms of water (e.g., ice) refer to volumes in millions of cubic kilometres; those following the processes (e.g., precipitation) refer to their fluxes in millions of cubic kilometres of water per year.

Carbon Dioxide

Of the greenhouse gases, carbon dioxide (CO_2) is the most significant. Natural sources of atmospheric CO_2 include outgassing from volcanoes, the combustion and natural decay of organic matter, and respiration by aerobic (oxygen-using) organisms. These sources are balanced, on average, by a set of physical, chemical, or biological processes, called "sinks," that tend to remove CO_2 from the atmosphere. Significant natural sinks include terrestrial vegetation, which takes up CO_2 during the process of photosynthesis.

The carbon cycleCarbon is transported in various forms through the atmosphere, the hydrosphere, and geologic formations. One of the primary pathways for the exchange of carbon dioxide (CO_2) takes place between the atmosphere and the oceans; there a fraction of the CO_2 combines with water, forming carbonic acid (H_2CO_3) that subsequently loses hydrogen ions (H^+) to form bicarbonate (HCO_3^-) and carbonate (CO_3^{2-}) ions. Mollusk shells or mineral precipitates that form by the reaction of calcium or other metal ions with carbonate may become buried in geologic strata and eventually release CO_2 through volcanic outgassing. Carbon dioxide also exchanges through photosynthesis in plants and through respiration in animals. Dead and decaying organic matter may ferment and release CO_2 or methane (CH_4) or may be incorporated into sedimentary rock, where it is converted to fossil fuels. Burning of hydrocarbon fuels returns CO_2 and water (H_2O) to the atmosphere. The biological and anthropogenic pathways are much faster than the geochemical pathways and, consequently, have a greater impact on the composition and temperature of the atmosphere.

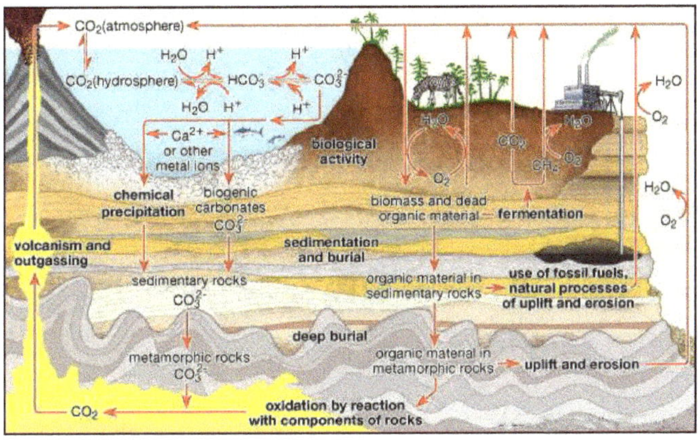

A number of oceanic processes also act as carbon sinks. One such process, called the "solubility pump," involves the descent of surface seawater containing dissolved CO_2. Another process, the "biological pump," involves the uptake of dissolved CO_2 by marine vegetation and phytoplankton (small free-floating photosynthetic organisms) living in the upper ocean or by other marine organisms that use CO_2 to build skeletons and other structures made of calcium carbonate ($CaCO_3$). As these organisms expire and fall to the ocean floor, the carbon they contain is transported downward and eventually buried at depth. A long-term balance between these natural sources and sinks leads to the background, or natural, level of CO_2 in the atmosphere.

In contrast, human activities increase atmospheric CO_2 levels primarily through the burning of fossil fuels—principally oil and coal and secondarily natural gas, for use in transportation, heating, and the generation of electrical power—and through the production of cement. Other anthropogenic sources include the burning of forests and the clearing of land. Anthropogenic emissions currently account for the annual release of about 7 gigatons (7 billion tons) of carbon into the atmosphere. Anthropogenic emissions are equal to approximately 3 percent of the total emissions of CO_2 by natural sources, and this amplified carbon load from human activities far exceeds the offsetting capacity of natural sinks (by perhaps as much as 2–3 gigatons per year).

Deforestation Smoldering remains of a plot of deforested land in the Amazon Rainforest of Brazil. Annually, it is estimated that net global deforestation accounts for about two gigatons of carbon emissions to the atmosphere.

CO_2 consequently accumulated in the atmosphere at an average rate of 1.4 ppm per year between 1959 and 2006 and roughly 2.0 ppm per year between 2006 and 2018. Overall, this rate of accumulation has been linear (that is, uniform over time). However, certain current sinks, such as the oceans, could become sources in the future. This may lead to a situation in which the concentration of atmospheric CO_2 builds at an exponential rate (that is, its rate of increase is also increasing).

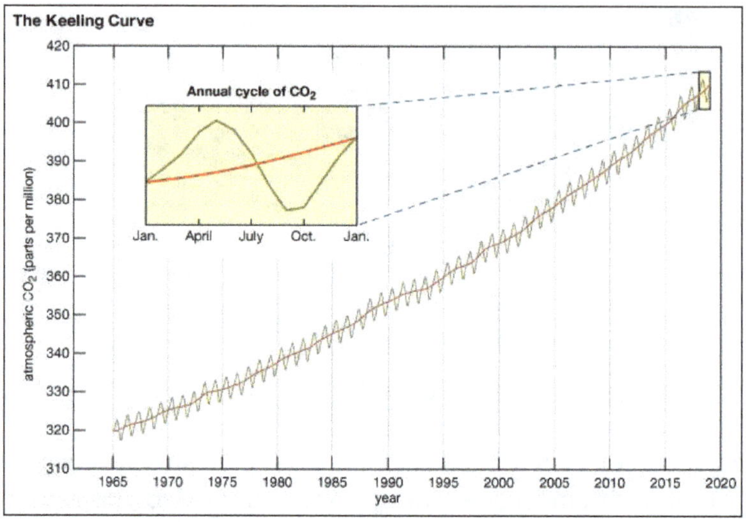

The Keeling Curve, named after American climate scientist Charles David Keeling, tracks changes in the concentration of carbon dioxide (CO_2) in Earth's atmosphere at a research station on Mauna Loa in Hawaii. Although these concentrations experience small seasonal fluctuations, the overall trend shows that CO_2 is increasing in the atmosphere.

The natural background level of carbon dioxide varies on timescales of millions of years because of slow changes in outgassing through volcanic activity. For example, roughly 100 million years ago, during the Cretaceous Period (145 million to 66 million years ago), CO_2 concentrations appear to have been several times higher than they are today (perhaps close to 2,000 ppm). Over the past 700,000 years, CO_2 concentrations have varied over a far smaller range (between roughly 180 and 300 ppm) in association with the same Earth orbital effects linked to the coming and going of the Pleistocene ice ages. By the early 21st century, CO_2 levels had reached 384 ppm, which is approximately 37 percent above the natural background level of roughly 280 ppm that existed at the beginning of the Industrial Revolution. Atmospheric CO_2 levels continued to increase, and by 2018 they had reached 410 ppm. Such levels are believed to be the highest in at least 800,000 years according to ice core measurements and may be the highest in at least 5 million years according to other lines of evidence.

Radiative forcing caused by carbon dioxide varies in an approximately logarithmic fashion with the concentration of that gas in the atmosphere. The logarithmic relationship occurs as the result of a saturation effect wherein it becomes increasingly difficult, as CO_2 concentrations increase, for additional CO_2 molecules to further influence the "infrared window" (a certain narrow band of wavelengths in the infrared region that is not absorbed by atmospheric gases). The logarithmic relationship predicts that the surface warming potential will rise by roughly the same amount for each doubling of CO_2 concentration. At current rates of fossil fuel use, a doubling of CO_2 concentrations over preindustrial levels is expected to take place by the middle of the 21st century (when CO_2 concentrations are projected to reach 560 ppm). A doubling of CO_2 concentrations would represent an increase of roughly 4 watts per square metre of radiative forcing. Given typical estimates of "climate sensitivity" in the absence of any offsetting factors, this energy increase would lead to a warming of 2 to 5 °C (3.6 to 9 °F) over preindustrial times. The total radiative forcing by anthropogenic CO_2 emissions since the beginning of the industrial age is approximately 1.66 watts per square metre.

Methane

Methane (CH_4) is the second most important greenhouse gas. CH_4 is more potent than CO_2 because the radiative forcing produced per molecule is greater. In addition, the infrared window is less saturated in the range of wavelengths of radiation absorbed by CH_4, so more molecules may fill in the region. However, CH_4 exists in far lower concentrations than CO_2 in the atmosphere, and its concentrations by volume in the atmosphere are generally measured in parts per billion (ppb) rather than ppm. CH_4 also has a considerably shorter residence time in the atmosphere than CO_2 (the residence time for CH_4 is roughly 10 years, compared with hundreds of years for CO_2).

Natural sources of methane include tropical and northern wetlands, methane-oxidizing bacteria that feed on organic material consumed by termites, volcanoes, seepage vents of the seafloor in regions rich with organic sediment, and methane hydrates trapped along the continental shelves of the oceans and in polar permafrost. The primary natural sink for methane is the atmosphere itself, as methane reacts readily with the hydroxyl radical (OH) within the troposphere to form CO_2 and water vapour (H_2O). When CH_4 reaches the stratosphere, it is destroyed. Another natural sink is soil, where methane is oxidized by bacteria.

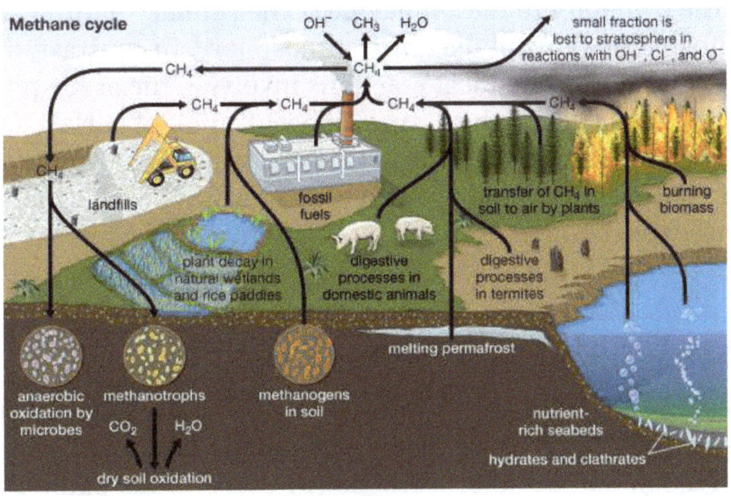

Methane cycle.

As with CO_2, human activity is increasing the CH_4 concentration faster than it can be offset by natural sinks. Anthropogenic sources currently account for approximately 70 percent of total annual emissions, leading to substantial increases in concentration over time. The major anthropogenic sources of atmospheric CH_4 are rice cultivation, livestock farming, the burning of coal and natural gas, the combustion of biomass, and the decomposition of organic matter in landfills. Future trends are particularly difficult to anticipate. This is in part due to an incomplete understanding of the climate feedbacks associated with CH_4 emissions. In addition it is difficult to predict how, as human populations grow, possible changes in livestock raising, rice cultivation, and energy utilization will influence CH_4 emissions.

It is believed that a sudden increase in the concentration of methane in the atmosphere was responsible for a warming event that raised average global temperatures by 4–8 °C (7.2–14.4 °F) over a few thousand years during the so-called Paleocene-Eocene Thermal Maximum, or PETM. This episode took place roughly 55 million years ago, and the rise in CH_4 appears to have been related to a massive volcanic eruption that interacted with methane-containing flood deposits. As a result, large amounts of gaseous CH_4 were injected into the atmosphere. It is difficult to know precisely how high these concentrations were or how long they persisted. At very high concentrations, residence times of CH_4 in the atmosphere can become much greater than the nominal 10-year residence time that applies today. Nevertheless, it is likely that these concentrations reached several ppm during the PETM.

Methane concentrations have also varied over a smaller range (between roughly 350 and 800 ppb) in association with the Pleistocene ice age cycles. Preindustrial levels of CH_4 in the atmosphere were approximately 700 ppb, whereas levels exceeded 1,867 ppb in late 2018. (These concentrations are well above the natural levels observed for at least the past 650,000 years.) The net radiative forcing by anthropogenic CH_4 emissions is approximately 0.5 watt per square metre—or roughly one-third the radiative forcing of CO_2.

Surface-Level Ozone and other Compounds

The next most significant greenhouse gas is surface, or low-level, ozone (O_3). Surface O_3 is a result of air pollution; it must be distinguished from naturally occurring stratospheric O_3, which has a very different role in the planetary radiation balance. The primary natural source of surface O_3 is the subsidence of stratospheric O_3 from the upper atmosphere. In contrast, the primary anthropogenic source of surface O_3 is photochemical reactions involving the atmospheric pollutant carbon monoxide (CO). The best estimates of the natural concentration of surface O_3 are 10 ppb, and the net radiative forcing due to anthropogenic emissions of surface O_3 is approximately 0.35 watt per square metre. Ozone concentrations can rise above unhealthy levels (that is, conditions where concentrations meet or exceed 70 ppb for eight hours or longer) in cities prone to photochemical smog.

Nitrous Oxides and Fluorinated Gases

Additional trace gases produced by industrial activity that have greenhouse properties include nitrous oxide (N_2O) and fluorinated gases (halocarbons), the latter including sulfur hexafluoride, hydrofluorocarbons (HFCs), and perfluorocarbons (PFCs). Nitrous oxide is responsible for 0.16 watt per square metre radiative forcing, while fluorinated gases are collectively responsible for 0.34 watt per square metre. Nitrous oxides have small background concentrations due to natural biological reactions in soil and water, whereas the fluorinated gases owe their existence almost entirely to industrial sources.

Aerosols

The production of aerosols represents an important anthropogenic radiative forcing of climate. Collectively, aerosols block—that is, reflect and absorb—a portion of incoming solar radiation, and this creates a negative radiative forcing. Aerosols are second only to greenhouse gases in relative importance in their impact on near-surface air temperatures. Unlike the decade-long residence times of the “well-mixed” greenhouse gases, such as CO_2 and CH_4, aerosols are readily flushed out of the atmosphere within days, either by rain or snow (wet deposition) or by settling out of the air (dry deposition). They must therefore be continually generated in order to produce a steady effect on radiative forcing. Aerosols have the ability to influence climate directly by absorbing or reflecting incoming solar radiation, but they can also produce indirect effects on climate by modifying cloud formation or cloud properties. Most aerosols serve as condensation nuclei (surfaces upon which water vapour can condense to form clouds); however, darker-coloured aerosols may hinder cloud formation by absorbing sunlight and heating up the surrounding air. Aerosols can be transported thousands of kilometres from their sources of origin by winds and upper-level circulation in the atmosphere.

Perhaps the most important type of anthropogenic aerosol in radiative forcing is sulfate aerosol. It is produced from sulfur dioxide (SO_2) emissions associated with the burning of coal and oil. Since the late 1980s, global emissions of SO_2 have decreased from about 151.5 million tonnes (167.0 million tons) to less than 100 million tonnes (110.2 million tons) of sulfur per year.

Nitrate aerosol is not as important as sulfate aerosol, but it has the potential to become a significant source of negative forcing. One major source of nitrate aerosol is smog (the combination of ozone with oxides of nitrogen in the lower atmosphere) released from the incomplete burning of fuel in internal-combustion engines. Another source is ammonia (NH_3), which is often used in fertilizers or released by the burning of plants and other organic materials. If greater amounts of atmospheric nitrogen are converted to ammonia and agricultural ammonia emissions continue to increase as projected, the influence of nitrate aerosols on radiative forcing is expected to grow.

Both sulfate and nitrate aerosols act primarily by reflecting incoming solar radiation, thereby reducing the amount of sunlight reaching the surface. Most aerosols, unlike greenhouse gases, impart a cooling rather than warming influence on Earth’s surface. One prominent exception is carbonaceous aerosols such as carbon black or soot, which are produced by the burning of fossil fuels and biomass. Carbon black tends to absorb rather than reflect incident solar radiation, and so it has a warming impact on the lower atmosphere, where it resides. Because of its absorptive properties, carbon black is also capable of having an additional indirect effect on climate. Through its deposition in snowfall, it can decrease the albedo of snow cover. This reduction in the amount of solar radiation reflected back to space by snow surfaces creates a minor positive radiative forcing.

Natural forms of aerosol include windblown mineral dust generated in arid and semiarid regions and sea salt produced by the action of waves breaking in the ocean. Changes to wind patterns as a result of climate modification could alter the emissions of these aerosols. The influence of climate change on regional patterns of aridity could shift both the sources and the destinations of dust clouds. In addition, since the concentration of sea salt aerosol, or sea aerosol, increases with the strength of the winds near the ocean surface, changes in wind speed due to global warming and climate change could influence the concentration of sea salt aerosol. For example, some studies suggest that climate change might lead to stronger winds over parts of the North Atlantic Ocean. Areas with stronger winds may experience an increase in the concentration of sea salt aerosol.

Other natural sources of aerosols include volcanic eruptions, which produce sulfate aerosol, and biogenic sources (e.g., phytoplankton), which produce dimethyl sulfide (DMS). Other important biogenic aerosols, such as terpenes, are produced naturally by certain kinds of trees or other plants.

For example, the dense forests of the Blue Ridge Mountains of Virginia in the United States emit terpenes during the summer months, which in turn interact with the high humidity and warm temperatures to produce a natural photochemical smog. Anthropogenic pollutants such as nitrate and ozone, both of which serve as precursor molecules for the generation of biogenic aerosol, appear to have increased the rate of production of these aerosols severalfold. This process appears to be responsible for some of the increased aerosol pollution in regions undergoing rapid urbanization.

Human activity has greatly increased the amount of aerosol in the atmosphere compared with the background levels of preindustrial times. In contrast to the global effects of greenhouse gases, the impact of anthropogenic aerosols is confined primarily to the Northern Hemisphere, where most of the world's industrial activity occurs. The pattern of increases in anthropogenic aerosol over time is also somewhat different from that of greenhouse gases. During the middle of the 20th century, there was a substantial increase in aerosol emissions. This appears to have been at least partially responsible for a cessation of surface warming that took place in the Northern Hemisphere from the 1940s through the 1970s. Since that time, aerosol emissions have leveled off due to antipollution measures undertaken in the industrialized countries since the 1960s. Aerosol emissions may rise in the future, however, as a result of the rapid emergence of coal-fired electric power generation in China and India.

The total radiative forcing of all anthropogenic aerosols is approximately −1.2 watts per square metre. Of this total, −0.5 watt per square metre comes from direct effects (such as the reflection of solar energy back into space), and −0.7 watt per square metre comes from indirect effects (such as the influence of aerosols on cloud formation). This negative radiative forcing represents an offset of roughly 40 percent from the positive radiative forcing caused by human activity. However, the relative uncertainty in aerosol radiative forcing (approximately 90 percent) is much greater than that of greenhouse gases. In addition, future emissions of aerosols from human activities, and the influence of these emissions on future climate change, are not known with any certainty. Nevertheless, it can be said that, if concentrations of anthropogenic aerosols continue to decrease as they have since the 1970s, a significant offset to the effects of greenhouse gases will be reduced, opening future climate to further warming.

Land-use Change

There are a number of ways in which changes in land use can influence climate. The most direct influence is through the alteration of Earth's albedo, or surface reflectance. For example, the replacement of forest by cropland and pasture in the middle latitudes over the past several centuries has led to an increase in albedo, which in turn has led to greater reflection of incoming solar radiation in those regions. This replacement of forest by agriculture has been associated with a change in global average radiative forcing of approximately −0.2 watt per square metre since 1750. In Europe and other major agricultural regions, such land-use conversion began more than 1,000 years ago and has proceeded nearly to completion. For Europe, the negative radiative forcing due to land-use change has probably been substantial, perhaps approaching −5 watts per square metre. The influence of early land use on radiative forcing may help to explain a long period of cooling in Europe that followed a period of relatively mild conditions roughly 1,000 years ago. It is generally believed that the mild temperatures of this "medieval warm period," which was followed by a long period of cooling, rivaled those of 20th-century Europe.

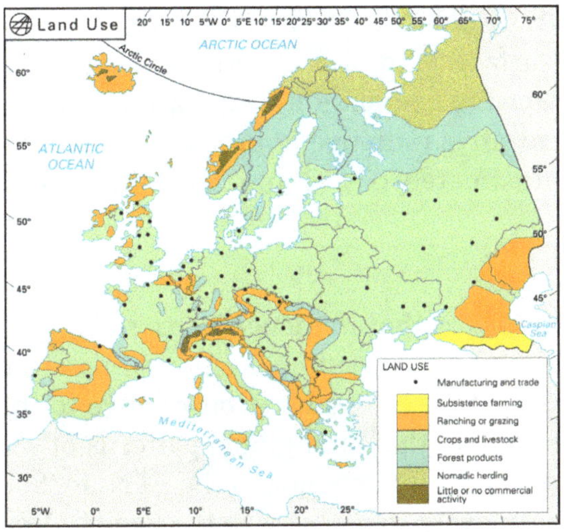

Europe: land useLand use in Europe.

Land-use changes can also influence climate through their influence on the exchange of heat between Earth's surface and the atmosphere. For example, vegetation helps to facilitate the evaporation of water into the atmosphere through evapotranspiration. In this process, plants take up liquid water from the soil through their root systems. Eventually this water is released through transpiration into the atmosphere, as water vapour through the stomata in leaves. While deforestation generally leads to surface cooling due to the albedo factor discussed above, the land surface may also be warmed as a result of the release of latent heat by the evapotranspiration process. The relative importance of these two factors, one exerting a cooling effect and the other a warming effect, varies by both season and region. While the albedo effect is likely to dominate in middle latitudes, especially during the period from autumn through spring, the evapotranspiration effect may dominate during the summer in the midlatitudes and year-round in the tropics. The latter case is particularly important in assessing the potential impacts of continued tropical deforestation.

The rate at which tropical regions are deforested is also relevant to the process of carbon sequestration, the long-term storage of carbon in underground cavities and biomass rather than in the atmosphere. By removing carbon from the atmosphere, carbon sequestration acts to mitigate global warming. Deforestation contributes to global warming, as fewer plants are available to take up carbon dioxide from the atmosphere. In addition, as fallen trees, shrubs, and other plants are burned or allowed to slowly decompose, they release as carbon dioxide the carbon they stored during their lifetimes. Furthermore, any land-use change that influences the amount, distribution, or type of vegetation in a region can affect the concentrations of biogenic aerosols, though the impact of such changes on climate is indirect and relatively minor.

Stratospheric Ozone Depletion

Since the 1970s the loss of ozone (O_3) from the stratosphere has led to a small amount of negative radiative forcing of the surface. This negative forcing represents a competition between two distinct effects caused by the fact that ozone absorbs solar radiation. In the first case, as ozone levels in the stratosphere are depleted, more solar radiation reaches Earth's surface. In the absence of any other influence, this rise in insolation would represent a positive radiative forcing of the surface. However, there is a second effect of ozone depletion that is related to its greenhouse properties. As the amount of ozone in the stratosphere is decreased, there is also less ozone to absorb longwave radiation emitted by Earth's surface. With less absorption of radiation by ozone, there is a corresponding decrease in the downward reemission of radiation. This second effect overwhelms the first and results in a modest negative radiative forcing of Earth's surface and a modest cooling of the lower stratosphere by approximately 0.5 °C (0.9 °F) per decade since the 1970s.

Natural Influences on Climate

There are a number of natural factors that influence Earth's climate. These factors include external influences such as explosive volcanic eruptions, natural variations in the output of the Sun, and slow changes in the configuration of Earth's orbit relative to the Sun. In addition, there are natural oscillations in Earth's climate that alter global patterns of wind circulation, precipitation, and surface temperatures. One such phenomenon is the El Niño/Southern Oscillation (ENSO), a coupled atmospheric and oceanic event that occurs in the Pacific Ocean every three to seven years. In addition, the Atlantic Multidecadal Oscillation (AMO) is a similar phenomenon that occurs over decades in the North Atlantic Ocean. Other types of oscillatory behaviour that produce dramatic shifts in climate may occur across timescales of centuries and millennia.

Volcanic Aerosols

A column of gas and ash rising from Mount Pinatubo in the Philippines on June 12, 1991, just days before the volcano's climactic explosion on June 15.

Explosive volcanic eruptions have the potential to inject substantial amounts of sulfate aerosols into the lower stratosphere. In contrast to aerosol emissions in the lower troposphere, aerosols that enter the stratosphere may remain for several years before settling out, because of the relative absence of turbulent motions there. Consequently, aerosols from explosive volcanic eruptions have the potential to affect Earth's climate. Less-explosive eruptions, or eruptions that are less vertical in orientation, have a lower potential for substantial climate impact. Furthermore, because of large-scale circulation patterns within the stratosphere, aerosols injected within tropical regions tend to spread out over the globe, whereas aerosols injected within midlatitude and polar regions tend to remain confined to the middle and high latitudes of that hemisphere. Tropical eruptions, therefore, tend to have a greater climatic impact than eruptions occurring toward the poles. In 1991 the moderate eruption of Mount Pinatubo in the Philippines provided a peak forcing of approximately −4 watts per square metre and cooled the climate by about 0.5 °C (0.9 °F) over the following few years. By comparison, the 1815 Mount Tambora eruption in present-day Indonesia, typically implicated for the 1816 "year without a summer" in Europe and North America, is believed to have been associated with a radiative forcing of approximately –6 watts per square metre.

While in the stratosphere, volcanic sulfate aerosol actually absorbs longwave radiation emitted by Earth's surface, and absorption in the stratosphere tends to result in a cooling of the troposphere below. This vertical pattern of temperature change in the atmosphere influences the behaviour of winds in the lower atmosphere, primarily in winter. Thus, while there is essentially a global cooling effect for the first few years following an explosive volcanic eruption, changes in the winter patterns of surface winds may actually lead to warmer winters in some areas, such as Europe. Some modern examples of major eruptions include Krakatoa (Indonesia) in 1883, El Chichón (Mexico) in 1982, and Mount Pinatubo in 1991. There is also evidence that volcanic eruptions may influence other climate phenomena such as ENSO.

Variations in Solar Output

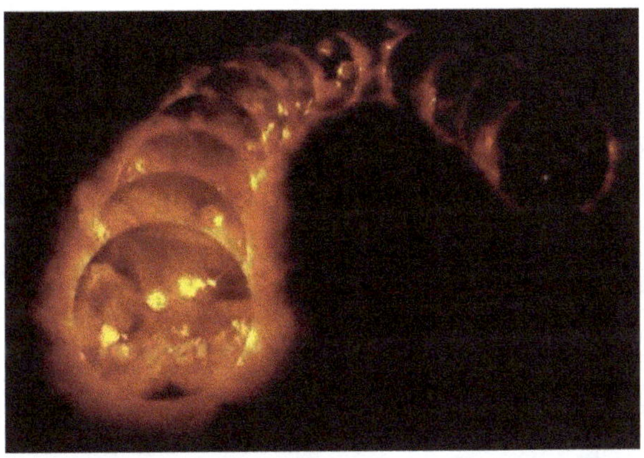

Twelve solar X-ray images obtained by Yohkoh between 1991 and 1995. The solar coronal brightness decreases by a factor of about 100 during a solar cycle as the Sun goes from an "active" state (left) to a less active state (right).

Direct measurements of solar irradiance, or solar output, have been available from satellites only since the late 1970s. These measurements show a very small peak-to-peak variation in solar irradiance (roughly 0.1 percent of the 1,366 watts per square metre received at the top of the atmosphere, for approximately 1.4 watts per square metre). However, indirect measures of solar activity are available from historical sunspot measurements dating back through the early 17th century. Attempts have been made to reconstruct graphs of solar irradiance variations from historical sunspot data by calibrating them against the measurements from modern satellites. However, since the modern measurements span only a few of the most recent 11-year solar cycles, estimates of solar output variability on 100-year and longer timescales are poorly correlated. Different assumptions regarding the relationship between the amplitudes of 11-year solar cycles and long-period solar output changes can lead to considerable differences in the resulting solar reconstructions. These differences in turn lead to fairly large uncertainty in estimating positive forcing by changes in solar irradiance since 1750. (Estimates range from 0.06 to 0.3 watt per square metre.) Even more challenging, given the lack of any modern analog, is the estimation of solar irradiance during the so-called Maunder Minimum, a period lasting from the mid-17th century to the early 18th century when very few sunspots were observed. While it is likely that solar irradiance was reduced at this time, it is difficult to calculate by how much. However, additional proxies of solar output exist that match reasonably well with the sunspot-derived records following the Maunder Minimum; these may be used as crude estimates of the solar irradiance variations.

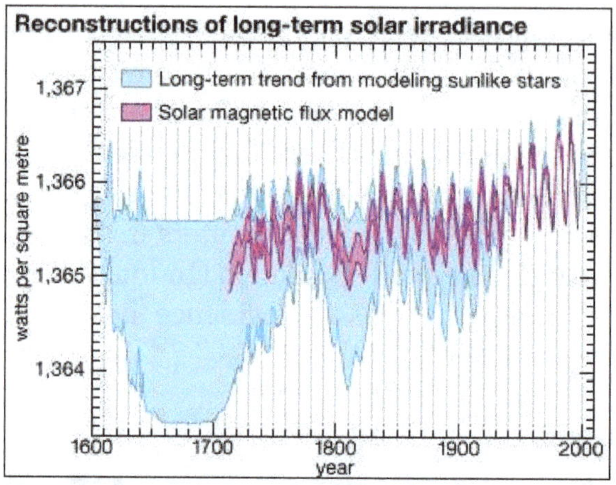

The trend shown in the longer reconstruction was inferred by Lean (2000) from modeling the changes in the brightness of stars similar to the Sun. The trend depicted in the shorter reconstruction by Y. Wang et al. (2005) was based on a magnetic flux model that simulated the long-term evolution of faculae (bright granular structures on the Sun's surface). Both models track a slight increase in solar irradiance since 1900.

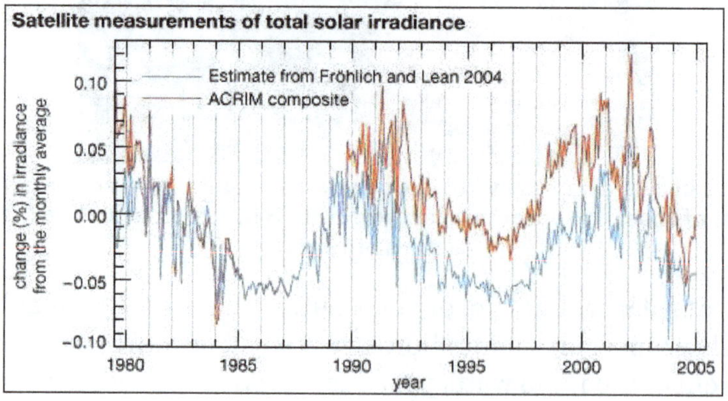

Monthly satellite measurements of total solar irradiance since 1980 comparing NASA's ACRIM-SAT data by Willson and Mordvinov (2003) with the Physikalisch-Meteorologisches Observatorium Davos (PMOD) composite developed by Fröhlich and Lean (2004). The PMOD composite combines ACRIM data collected by the Solar Maximum Mission (SMM) and Upper Atmosphere Research Satellite (UARS) with those provided by the Solar and Heliospheric Observatory (SOHO) and Nimbus 7 satellites.

In theory it is possible to estimate solar irradiance even farther back in time, over at least the past millennium, by measuring levels of cosmogenic isotopes such as carbon-14 and beryllium-10. Cosmogenic isotopes are isotopes that are formed by interactions of cosmic rays with atomic nuclei in the atmosphere and that subsequently fall to Earth, where they can be measured in the annual layers found in ice cores. Since their production rate in the upper atmosphere is modulated by changes in solar activity, cosmogenic isotopes may be used as indirect indicators of solar irradiance. However, as with the sunspot data, there is still considerable uncertainty in the amplitude of past solar variability implied by these data.

Solar forcing also affects the photochemical reactions that manufacture ozone in the stratosphere. Through this modulation of stratospheric ozone concentrations, changes in solar irradiance (particularly in the ultraviolet portion of the electromagnetic spectrum) can modify how both shortwave and longwave radiation in the lower stratosphere are absorbed. As a result, the vertical temperature profile of the atmosphere can change, and this change can in turn influence phenomena such as the strength of the winter jet streams.

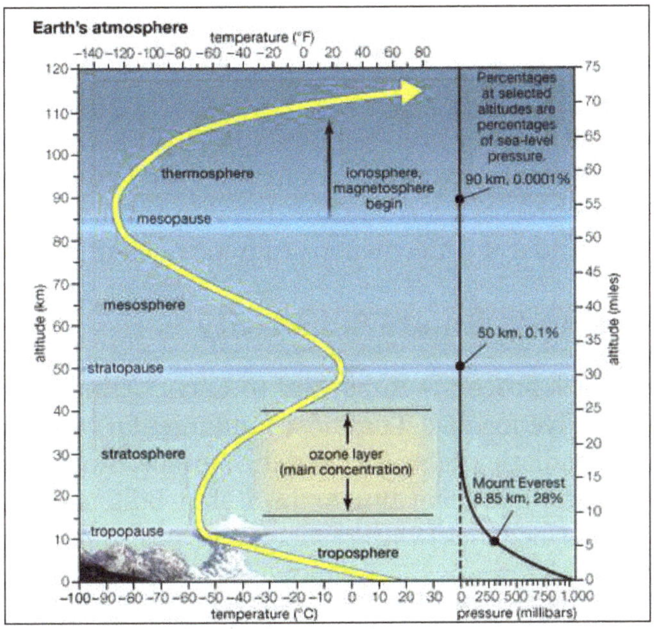

The layers of Earth's atmosphere. The yellow line shows the response of air temperature to increasing height.

Variations in Earth's Orbit

On timescales of tens of millennia, the dominant radiative forcing of Earth's climate is associated with slow variations in the geometry of Earth's orbit about the Sun. These variations include the precession of the equinoxes (that is, changes in the timing of summer and winter), occurring on a roughly 26,000-year timescale; changes in the tilt angle of Earth's rotational axis relative to the plane of Earth's orbit around the Sun, occurring on a roughly 41,000-year timescale; and changes in the eccentricity (the departure from a perfect circle) of Earth's orbit around the Sun, occurring on a roughly 100,000-year timescale. Changes in eccentricity slightly influence the mean annual solar radiation at the top of Earth's atmosphere, but the primary influence of all the orbital variations listed above is on the seasonal and latitudinal distribution of incoming solar radiation over Earth's surface. The major ice ages of the Pleistocene Epoch were closely related to the influence of these variations on summer insolation at high northern latitudes. Orbital variations thus exerted a primary control on the extent of continental ice sheets. However, Earth's orbital changes are generally believed to have had little impact on climate over the past few millennia, and so they are not considered to be significant factors in present-day climate variability.

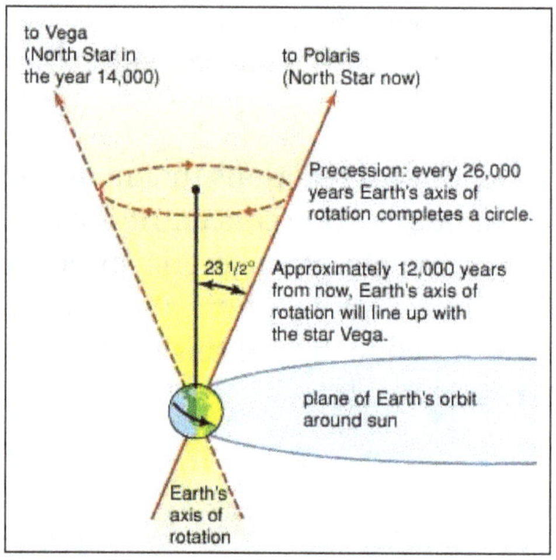

Earth's axis of rotation itself rotates, or precesses, completing one circle every 26,000 years. Consequently, Earth's North Pole points toward different stars (and sometimes toward empty space) as it travels in this circle. This precession is so slow that it is not noticeable in a person's lifetime, though astronomers must consider its effect when studying ancient sites such as Stonehenge.

Feedback Mechanisms and Climate Sensitivity

There are a number of feedback processes important to Earth's climate system and, in particular, its response to external radiative forcing. The most fundamental of these feedback mechanisms involves the loss of longwave radiation to space from the surface. Since this radiative loss increases with increasing surface temperatures according to the Stefan-Boltzmann law, it represents a stabilizing factor (that is, a negative feedback) with respect to near-surface air temperature.

Climate sensitivity can be defined as the amount of surface warming resulting from each additional watt per square metre of radiative forcing. Alternatively, it is sometimes defined as the warming that would result from a doubling of CO_2 concentrations and the associated addition of 4 watts per square metre of radiative forcing. In the absence of any additional feedbacks, climate sensitivity would be approximately 0.25 °C (0.45 °F) for each additional watt per square metre of radiative forcing. Stated alternatively, if the CO_2 concentration of the atmosphere present at the start of the industrial age (280 ppm) were doubled (to 560 ppm), the resulting additional 4 watts per square metre of radiative forcing would translate into a 1 °C (1.8 °F) increase in air temperature. However, there are additional feedbacks that exert a destabilizing, rather than stabilizing, influence, and these feedbacks tend to increase the sensitivity of climate to somewhere between 0.5 and 1.0 °C (0.9 and 1.8 °F) for each additional watt per square metre of radiative forcing.

Water Vapour Feedback

Unlike concentrations of other greenhouse gases, the concentration of water vapour in the atmosphere cannot freely vary. Instead, it is determined by the temperature of the lower atmosphere and surface through a physical relationship known as the Clausius-Clapeyron equation, named for 19th-century German physicist Rudolf Clausius and 19th-century French engineer Émile Clapeyron. Under the assumption that there is a liquid water surface in equilibrium with the atmosphere, this relationship indicates that an increase in the capacity of air to hold water vapour is a function of increasing temperature of that volume of air. This assumption is relatively good over the oceans, where water is plentiful, but not over the continents. For this reason the relative humidity (the percent of water vapour the air contains relative to its capacity) is approximately 100 percent over ocean regions and much lower over continental regions (approaching 0 percent in arid regions). Not surprisingly, the average relative humidity of Earth's lower atmosphere is similar to the fraction of Earth's surface covered by the oceans (that is, roughly 70 percent). This quantity is expected to remain approximately constant as Earth warms or cools. Slight changes to global relative humidity may result from human land-use modification, such as tropical deforestation and irrigation, which can affect the relative humidity over land areas up to regional scales.

The amount of water vapour in the atmosphere will rise as the temperature of the atmosphere rises. Since water vapour is a very potent greenhouse gas, even more potent than CO_2, the net greenhouse effect actually becomes stronger as the surface warms, which leads to even greater warming. This positive feedback is known as the "water vapour feedback." It is the primary reason that climate sensitivity is substantially greater than the previously stated theoretical value of 0.25 °C (0.45 °F) for each increase of 1 watt per square metre of radiative forcing.

Cloud Feedbacks

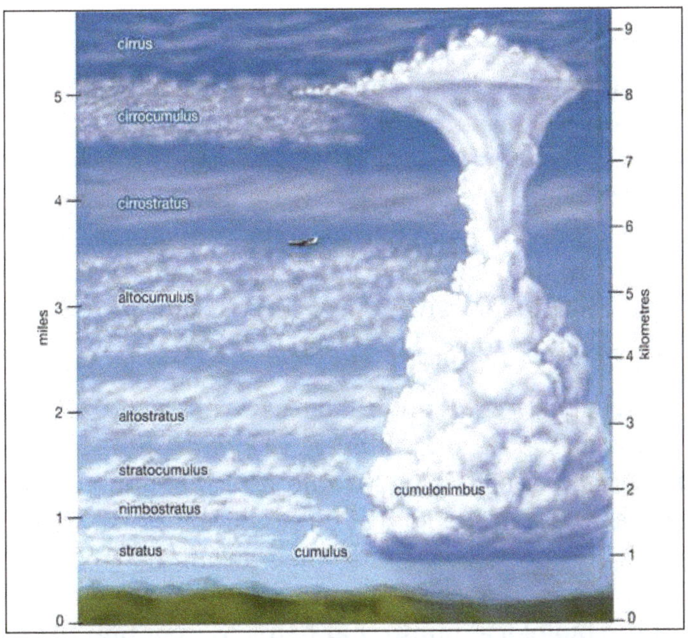

Different types of clouds form at different heights.

It is generally believed that as Earth's surface warms and the atmosphere's water vapour content increases, global cloud cover increases. However, the effects on near-surface air temperatures are complicated. In the case of low clouds, such as marine stratus clouds, the dominant radiative feature of the cloud is its albedo. Here any increase in low cloud cover acts in much the same way as an increase in surface ice cover: more incoming solar radiation is reflected and Earth's surface cools. On the other hand, high clouds, such as the towering cumulus clouds that extend up to the boundary between the troposphere and stratosphere, have a quite different impact on the surface radiation balance. The tops of cumulus clouds are considerably higher in the atmosphere and colder than their undersides. Cumulus cloud tops emit less longwave radiation out to space than the warmer cloud bottoms emit downward toward the surface. The end result of the formation of high cumulus clouds is greater warming at the surface.

The net feedback of clouds on rising surface temperatures is therefore somewhat uncertain. It represents a competition between the impacts of high and low clouds, and the balance is difficult to determine. Nonetheless, most estimates indicate that clouds on the whole represent a positive feedback and thus additional warming.

Ice Albedo Feedback

Another important positive climate feedback is the so-called ice albedo feedback. This feedback arises from the simple fact that ice is more reflective (that is, has a higher albedo) than land or water surfaces. Therefore, as global ice cover decreases, the reflectivity of Earth's surface decreases, more incoming solar radiation is absorbed by the surface, and the surface warms. This feedback is considerably more important when there is relatively extensive global ice cover, such as during the height of the last ice age, roughly 25,000 years ago. On a global scale the importance of ice albedo feedback decreases as Earth's surface warms and there is relatively less ice available to be melted.

Carbon Cycle Feedbacks

Another important set of climate feedbacks involves the global carbon cycle. In particular, the two main reservoirs of carbon in the climate system are the oceans and the terrestrial biosphere. These reservoirs have historically taken up large amounts of anthropogenic CO_2 emissions. Roughly 50–70 percent is removed by the oceans, whereas the remainder is taken up by the terrestrial biosphere. Global warming, however, could decrease the capacity of these reservoirs to sequester atmospheric CO_2. Reductions in the rate of carbon uptake by these reservoirs would increase the pace of CO_2 buildup in the atmosphere and represent yet another possible positive feedback to increased greenhouse gas concentrations.

In the world's oceans, this feedback effect might take several paths. First, as surface waters warm, they would hold less dissolved CO_2. Second, if more CO_2 were added to the atmosphere and taken up by the oceans, bicarbonate ions (HCO_3^-) would multiply and ocean acidity would increase. Since calcium carbonate ($CaCO_3$) is broken down by acidic solutions, rising acidity would threaten ocean-dwelling fauna that incorporate $CaCO_3$ into their skeletons or shells. As it becomes increasingly difficult for these organisms to absorb oceanic carbon, there would be a corresponding decrease in the efficiency of the biological pump that helps to maintain the oceans as a carbon sink. Third, rising surface temperatures might lead to a slowdown in the so-called thermohaline circulation, a global pattern of oceanic flow that partly drives the sinking of surface waters near the poles and is responsible for much of the burial of carbon in the deep ocean. A slowdown in this flow due to an influx of melting fresh water into what are normally saltwater conditions might also cause the solubility pump, which transfers CO_2 from shallow to deeper waters, to become less efficient. Indeed, it is predicted that if global warming continued to a certain point, the oceans would cease to be a net sink of CO_2 and would become a net source.

As large sections of tropical forest are lost because of the warming and drying of regions such as Amazonia, the overall capacity of plants to sequester atmospheric CO_2 would be reduced. As a result, the terrestrial biosphere, though currently a carbon sink, would become a carbon source. Ambient temperature is a significant factor affecting the pace of photosynthesis in plants, and many plant species that are well adapted to their local climatic conditions have maximized their photosynthetic rates. As temperatures increase and conditions begin to exceed the optimal temperature range for both photosynthesis and soil respiration, the rate of photosynthesis would decline. As dead plants decompose, microbial metabolic activity (a CO_2 source) would increase and would eventually outpace photosynthesis.

Under sufficient global warming conditions, methane sinks in the oceans and terrestrial biosphere also might become methane sources. Annual emissions of methane by wetlands might either increase or decrease, depending on temperatures and input of nutrients, and it is possible that wetlands could switch from source to sink. There is also the potential for increased methane release as a result of the warming of Arctic permafrost (on land) and further methane release at the continental margins of the oceans (a few hundred metres below sea level). The current average atmospheric methane concentration of 1,750 ppb is equivalent to 3.5 gigatons (3.5 billion tons) of carbon. There are at least 400 gigatons of carbon equivalent stored in Arctic permafrost and as much as 10,000 gigatons (10 trillion tons) of carbon equivalent trapped on the continental margins of the oceans in a hydrated crystalline form known as clathrate. It is believed that some fraction of this trapped methane could become unstable with additional warming, although the amount and rate of potential emission remain highly uncertain.

Climate Research

Modern research into climatic variation and change is based on a variety of empirical and theoretical lines of inquiry. One line of inquiry is the analysis of data that record changes in atmosphere, oceans, and climate from roughly 1850 to the present. In a second line of inquiry, information describing paleoclimatic changes is gathered from “proxy,” or indirect, sources such as ocean and lake sediments, pollen grains, corals, ice cores, and tree rings. Finally, a variety of theoretical models can be used to investigate the behaviour of Earth’s climate under different conditions.

Modern Observations

Although a limited regional subset of land-based records is available from the 17th and 18th centuries, instrumental measurements of key climate variables have been collected systematically and at global scales since the mid-19th to early 20th century. These data include measurements of surface temperature on land and at sea, atmospheric pressure at sea level, precipitation over continents and oceans, sea ice extents, surface winds, humidity, and tides. Such records are the most reliable of all available climate data, since they are precisely dated and are based on well-understood instruments and physical principles. Corrections must be made for uncertainties in the data (for instance, gaps in the observational record, particularly during earlier years) and for systematic errors (such as an “urban heat island” bias in temperature measurements made on land).

Since the mid-20th century a variety of upper-air observations have become available (for example, of temperature, humidity, and winds), allowing climatic conditions to be characterized from the ground upward through the upper troposphere and lower stratosphere. Since the 1970s these data have been supplemented by polar-orbiting and geostationary satellites and by platforms in the oceans that gauge temperature, salinity, and other properties of seawater. Attempts have been made to fill the gaps in early measurements by using various statistical techniques and "backward prediction" models and by assimilating available observations into numerical weather prediction models. These techniques seek to estimate meteorological observations or atmospheric variables (such as relative humidity) that have been poorly measured in the past.

Modern measurements of greenhouse gas concentrations began with an investigation of atmospheric carbon dioxide (CO_2) concentrations by American climate scientist Charles Keeling at the summit of Mauna Loa in Hawaii in 1958. Keeling's findings indicated that CO_2 concentrations were steadily rising in association with the combustion of fossil fuels, and they also yielded the famous "Keeling curve," a graph in which the longer-term rising trend is superimposed on small oscillations related to seasonal variations in the uptake and release of CO_2 from photosynthesis and respiration in the terrestrial biosphere. Keeling's measurements at Mauna Loa apply primarily to the Northern Hemisphere.

Taking into account the uncertainties, the instrumental climate record indicates substantial trends since the end of the 19th century consistent with a warming Earth. These trends include a rise in global surface temperature of 0.9 °C (1.5 °F) between 1880 and 2012, an associated elevation of global sea level of 19–21 cm (7.5–8.3 inches) between 1901 and 2010, and a decrease in snow cover in the Northern Hemisphere of approximately 1.5 million square km (580,000 square miles). Records of average global temperatures kept by the World Meteorological Organization (WMO) indicate that the years 1998, 2005, and 2010 are statistically tied with one another as the warmest years since modern record keeping began in 1880; the WMO also noted that the decade 2001–10 was the warmest decade since 1880. Increases in global sea level are attributed to a combination of seawater expansion due to ocean heating and freshwater runoff caused by the melting of terrestrial ice. Reductions in snow cover are the result of warmer temperatures favouring a steadily shrinking winter season.

Climate data collected during the first two decades of the 21st century reveal that surface warming between 2005 and 2014 proceeded slightly more slowly than was expected from the effect of greenhouse gas increases alone. This fact was sometimes used to suggest that global warming had stopped or that it experienced a "hiatus" or "pause." In reality, this phenomenon appears to have been influenced by several factors, none of which, however, implies that global warming stopped during this period or that global warming would not continue in the future. One factor was the increased burial of heat beneath the ocean surface by strong trade winds, a process assisted by La Niña conditions. The effects of La Niña manifest in the form of cooling surface waters along the western coast of South America. As a result, warming at the ocean surface was reduced, but the accumulation of heat in other parts of the ocean occurred at an accelerated rate. Another factor cited by climatologists was a small but potentially important increase in aerosols from volcanic activity, which may have blocked a small portion of incoming solar radiation and which were accompanied by a small reduction in solar output during the period. These factors, along with natural decades-long oscillations in the climate system, may have masked a portion of the greenhouse warming. (However, climatologists point out that these natural climate cycles are expected to add to greenhouse warming in the future when the oscillations eventually reverse direction.) For these reasons many scientists believe that it is an error to call this slowdown in detectable surface warming a "hiatus" or a "pause."

Prehistorical Climate Records

In order to reconstruct climate changes that occurred prior to about the mid-19th century, it is necessary to use "proxy" measurements—that is, records of other natural phenomena that indirectly measure various climate conditions. Some proxies, such as most sediment cores and pollen records, glacial moraine evidence, and geothermal borehole temperature profiles, are coarsely resolved or dated and thus are only useful for describing climate changes on long timescales. Other proxies, such as growth rings from trees or oxygen isotopes from corals and ice cores, can provide a record of yearly or even seasonal climate changes.

The data from these proxies should be calibrated to known physical principles or related statistically to the records collected by modern instruments, such as satellites. Networks of proxy data can then be used to infer patterns of change in climate variables, such as the behaviour of surface temperature over time and geography. Yearly reconstructions of climate variables are possible over the past 1,000 to 2,000 years using annually dated proxy records, but reconstructions farther back in time are generally based on more coarsely resolved evidence such as ocean sediments and pollen records. For these, records of conditions can be reconstructed only on timescales of hundreds or thousands of years. In addition, since relatively few long-term proxy records are available for the Southern Hemisphere, most reconstructions focus on the Northern Hemisphere.

The various proxy-based reconstructions of the average surface temperature of the Northern Hemisphere differ in their details. These differences are the result of uncertainties implicit in the proxy data themselves and also of differences in the statistical methods used to relate the proxy data to surface temperature. Nevertheless, all studies as reviewed in the IPCC's Fourth Assessment Report (AR4), which was published in 2007, indicate that the average surface temperature since about 1950 is higher than at any time during the previous 1,000 years.

Theoretical Climate Models

Theoretical models of Earth's climate system can be used to investigate the response of climate to external radiative forcing as well as its own internal variability. Two or more models that focus on different physical processes may be coupled or linked together through a common feature, such as geographic location. Climate models vary considerably in their degree of complexity. The simplest models of energy balance describe Earth's surface as a globally uniform layer whose temperature is determined by a balance of incoming and outgoing shortwave and longwave radiation. These simple models may also consider the effects of greenhouse gases. At the other end of the spectrum are fully coupled, three-dimensional, global climate models. These are complex models that solve for radiative balance; for laws of motion governing the atmosphere, ocean, and ice; and for exchanges of energy and momentum within and between the different components of the climate. In some cases, theoretical climate models also include an interactive representation of Earth's biosphere and carbon cycle.

Even the most-detailed climate models cannot resolve all the processes that are important in the atmosphere and ocean. Most climate models are designed to gauge the behaviour of a number of physical variables over space and time, and they often artificially divide Earth's surface into a grid of many equal-sized "cells." Each cell may neatly correspond to some physical process (such as summer near-surface air temperature) or other variable (such as land-use type), and it may be assigned a relatively straightforward value. So-called "sub-grid-scale" processes, such as those of clouds, are too small to be captured by the relatively coarse spacing of the individual grid cells. Instead, such processes must be represented through a statistical process that relates the properties of the atmosphere and ocean. For example, the average fraction of cloud cover over a hypothetical "grid box" (that is, a representative volume of air or water in the model) can be estimated from the average relative humidity and the vertical temperature profile of the grid cell. Variations in the behaviour of different coupled climate models arise in large part from differences in the ways sub-grid-scale processes are mathematically expressed.

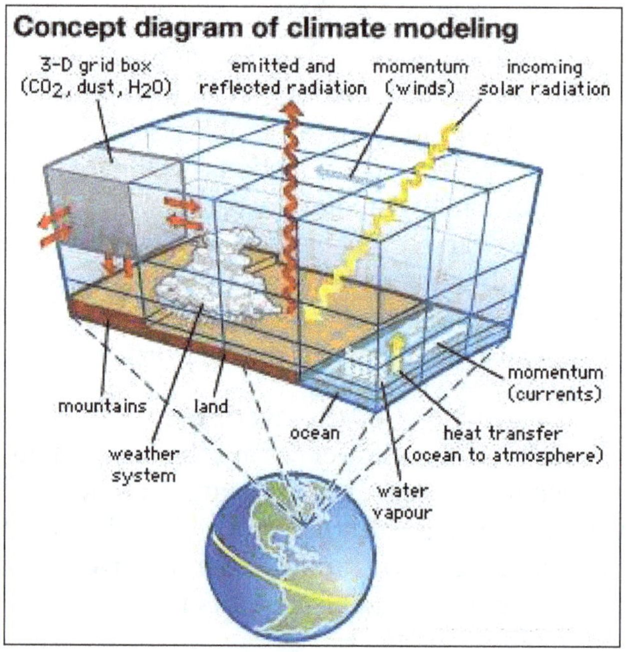

To understand and explain the complex behaviour of Earth's climate, modern climate models incorporate several variables that stand in for materials passing through Earth's atmosphere and oceans and the forces that affect them.

Despite these required simplifications, many theoretical climate models perform remarkably well when reproducing basic features of the atmosphere, such as the behaviour of midlatitude jet streams or Hadley cell circulation. The models also adequately reproduce important features of the oceans, such as the Gulf Stream. In addition, models are becoming better able to reproduce the main patterns of internal climate variability, such as those of El Niño/Southern Oscillation (ENSO). Consequently, periodically recurring events—such as ENSO and other interactions between the atmosphere and ocean currents—are being modeled with growing confidence.

Climate models have been tested in their ability to reproduce observed changes in response to radiative forcing. In 1988 a team at NASA's Goddard Institute for Space Studies in New York City used a fairly primitive climate model to predict warming patterns that might occur in response to three different scenarios of anthropogenic radiative forcing. Warming patterns were forecast for subsequent decades. Of the three scenarios, the middle one, which corresponds most closely to actual historical carbon emissions, comes closest to matching the observed warming of roughly 0.5 °C (0.9 °F) that has taken place since then. The NASA team also used a climate model to successfully predict that global mean surface temperatures would cool by about 0.5 °C for one to two years after the 1991 eruption of Mount Pinatubo in the Philippines.

More recently, so-called "detection and attribution" studies have been performed. These studies compare predicted changes in near-surface air temperature and other climate variables with patterns of change that have been observed for the past one to two centuries. The simulations have shown that the observed patterns of warming of Earth's surface and upper oceans, as well as changes in other climate phenomena such as prevailing winds and precipitation patterns, are consistent with the effects of an anthropogenic influence predicted by the climate models. In addition, climate model simulations have shown success in reproducing the magnitude and the spatial pattern of cooling in the Northern Hemisphere between roughly 1400 and 1850—during the Little Ice Age, which appears to have resulted from a combination of lowered solar output and heightened explosive volcanic activity.

Potential effects of Global Warming

The path of future climate change will depend on what courses of action are taken by society—in particular the emission of greenhouse gases from the burning of fossil fuels. A range of alternative emissions scenarios known as representative concentration pathways (RCPs) were proposed by the IPCC in the Fifth Assessment Report (AR5), which was published in 2014, to examine potential future climate changes. The scenarios depend on various assumptions concerning future rates of human population growth, economic development, energy demand, technological advancement, and other factors. Unlike the scenarios used in previous IPCC assessments, the AR5 RCPs explicitly account for climate change mitigation efforts.

Table: Projected range of sea-level rise by climate change scenario.

Scenario	temperature change (°C) in 2090–99 relative to 1980–99	sea-level rise (m) in 2090–99 relative to 1980–99
B1	1.1–2.9	0.18–0.38
A1T	1.4–3.8	0.20–0.45
B2	1.4–3.8	0.20–0.43
A1B	1.7–4.4	0.21–0.48
A2	2.0–5.4	0.23–0.51
A1Fl	2.4–6.4	0.26–0.59

The results of each scenario in the IPCC's Fourth Assessment Report (2007) are depicted in the graph.

The AR5 scenario with the smallest increases in greenhouse gases is RCP 2.6, which denotes the net radiative forcing by 2100 in watts per square metre (a doubling of CO_2 concentrations from preindustrial values of 280 ppm to 560 ppm represents roughly 3.7 watts per square metre). RCP 2.6 assumes substantial improvements in energy efficiency, a rapid transition away from fossil fuel energy, and a global population that peaks at roughly nine billion people in the 21st century. In that scenario CO_2 concentrations remain below 450 ppm and actually fall toward the end of the century (to about 420 ppm) as a result of widespread deployment of carbon-capture technology.

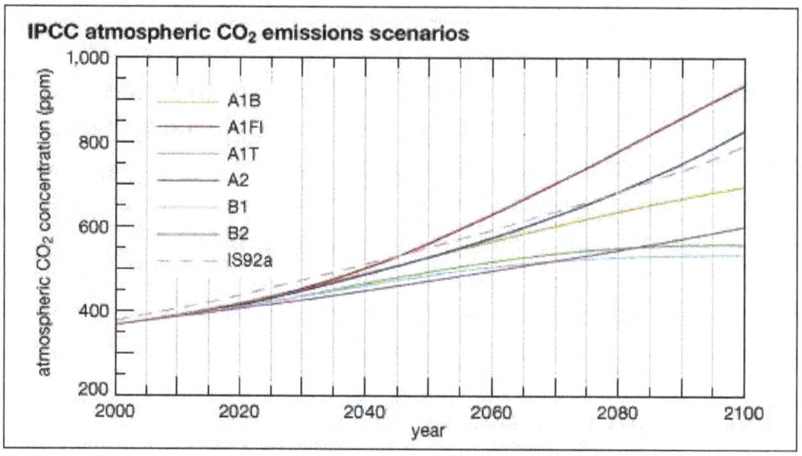

Carbon dioxide: Global warming scenarios Graph of the predicted increase in the concentration of carbon dioxide (CO_2) in Earth's atmosphere according to a series of climate change scenarios that assume different levels of economic development, population growth, and fossil fuel use.

Scenario RCP 8.5, by contrast, might be described as "business as usual." It reflects the assumption of an energy-intensive global economy, high population growth, and a reduced rate of technological development. CO_2 concentrations are more than three times greater than preindustrial levels (roughly 936 ppm) by 2100 and continue to grow thereafter. RCP 4.5 and RCP 6.0 envision intermediate policy choices, resulting in stabilization by 2100 of CO_2 concentrations at 538 and 670 ppm, respectively. In all those scenarios, the cooling effect of industrial pollutants such as sulfate particulates, which have masked some of the past century's warming, is assumed to decline to near zero by 2100 because of policies restricting their industrial production.

Simulations of Future Climate Change

The differences between the various simulations arise from disparities between the various climate models used and from assumptions made by each emission scenario. For example, best estimates of the predicted increases in global surface temperature between the years 2000 and 2100 range from about 0.3 to 4.8 °C (0.5 to 8.6 °F), depending on which emission scenario is assumed and which climate model is used. Relative to preindustrial (i.e., 1750–1800) temperatures, these estimates reflect an overall warming of the globe of 1.4 to 5.0 °C (2.5 to 9.0 °F). These projections are conservative in that they do not take into account potential positive carbon cycle feedbacks. Only the lower-end emissions scenario RCP 2.6 has a reasonable chance (roughly 50 percent) of holding additional global surface warming by 2100 to less than 2.0 °C (3.6 °F)—a level considered by many scientists to be the threshold above which pervasive and extreme climatic effects will occur.

Patterns of Warming

The greatest increase in near-surface air temperature is projected to occur over the polar region of the Northern Hemisphere because of the melting of sea ice and the associated reduction in surface albedo. Greater warming is predicted over land areas than over the ocean. Largely due to the delayed warming of the oceans and their greater specific heat, the Northern Hemisphere—with less than 40 percent of its surface area covered by water—is expected to warm faster than the Southern Hemisphere. Some of the regional variation in predicted warming is expected to arise from changes to wind patterns and ocean currents in response to surface warming. For example, the warming of the region of the North Atlantic Ocean just south of Greenland is expected to be slight. This anomaly is projected to arise from a weakening of warm northward ocean currents combined with a shift in the jet stream that will bring colder polar air masses to the region.

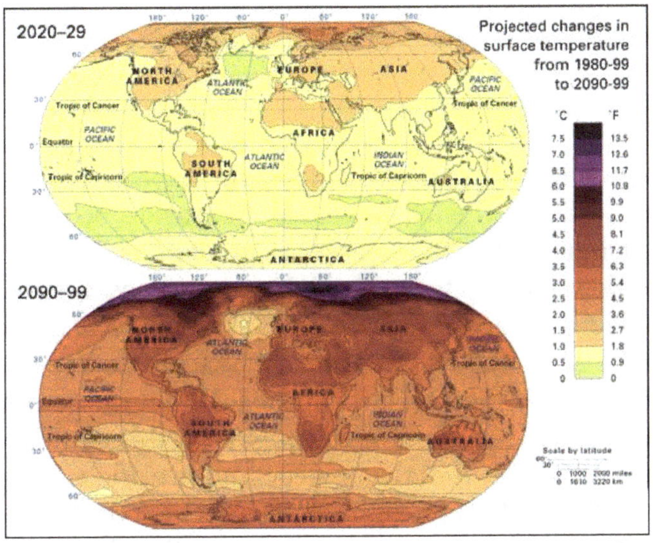

Projected changes in mean surface temperatures by the late 21st century according to the A1B climate change scenario. All values for the period 2090–99 are shown relative to the mean temperature values for the period 1980–99.

Precipitation Patterns

The climate changes associated with global warming are also projected to lead to changes in precipitation patterns across the globe. Increased precipitation is predicted in the polar and subpolar regions, whereas decreased precipitation is projected for the middle latitudes of both hemispheres as a result of the expected poleward shift in the jet streams. Whereas precipitation near the Equator is predicted to increase, it is thought that rainfall in the subtropics will decrease. Both phenomena are associated with a forecasted strengthening of the tropical Hadley cell pattern of atmospheric circulation.

Changes in precipitation patterns are expected to increase the chances of both drought and flood conditions in many areas. Decreased summer precipitation in North America, Europe, and Africa, combined with greater rates of evaporation due to warming surface temperatures, is projected to lead to decreased soil moisture and drought in many regions. Furthermore, since anthropogenic climate change will likely lead to a more vigorous hydrologic cycle with greater rates of both evaporation and precipitation, there will be a greater probability for intense precipitation and flooding in many regions.

Regional Predictions

Regional predictions of future climate change remain limited by uncertainties in how the precise patterns of atmospheric winds and ocean currents will vary with increased surface warming. For example, some uncertainty remains in how the frequency and magnitude of El Niño/Southern Oscillation (ENSO) events will adjust to climate change. Since ENSO is one of the most prominent sources of interannual variations in regional patterns of precipitation and temperature, any uncertainty in how it will change implies a corresponding uncertainty in certain regional patterns of climate change. For example, increased El Niño activity would likely lead to more winter precipitation in some regions, such as the desert southwest of the United States. This might offset the drought predicted for those regions, but at the same time it might lead to less precipitation in other regions. Rising winter precipitation in the desert southwest of the United States might exacerbate drought conditions in locations as far away as South Africa.

Ice Melt and Sea Level Rise

A warming climate holds important implications for other aspects of the global environment. Because of the slow process of heat diffusion in water, the world's oceans are likely to continue to warm for several centuries in response to increases in greenhouse concentrations that have taken place so far. The combination of seawater's thermal expansion associated with this warming and the melting of mountain glaciers is predicted to lead to an increase in global sea level of 0.45–0.82 metre (1.4–2.7 feet) by 2100 under the RCP 8.5 emissions scenario. However, the actual rise in sea level could be considerably greater than this. It is probable that the continued warming of Greenland will cause its ice sheet to melt at accelerated rates. In addition, this level of surface warming may also melt the ice sheet of West Antarctica. Paleoclimatic evidence suggests that an additional 2 °C (3.6 °F) of warming could lead to the ultimate destruction of the Greenland Ice Sheet, an event that would add another 5 to 6 metres (16 to 20 feet) to predicted sea level rise. Such an increase would submerge a substantial number of islands and lowland regions. Coastal lowland regions vulnerable to sea level rise include substantial parts of the U.S. Gulf Coast and Eastern Seaboard (including roughly the lower third of Florida), much of the Netherlands and Belgium (two of the European Low Countries), and heavily populated tropical areas such as Bangladesh. In addition, many of the world's major cities—such as Tokyo, New York, Mumbai, Shanghai, and Dhaka—are located in lowland regions vulnerable to rising sea levels. With the loss of the West Antarctic ice sheet, additional sea level rise would approach 10.5 metres (34 feet).

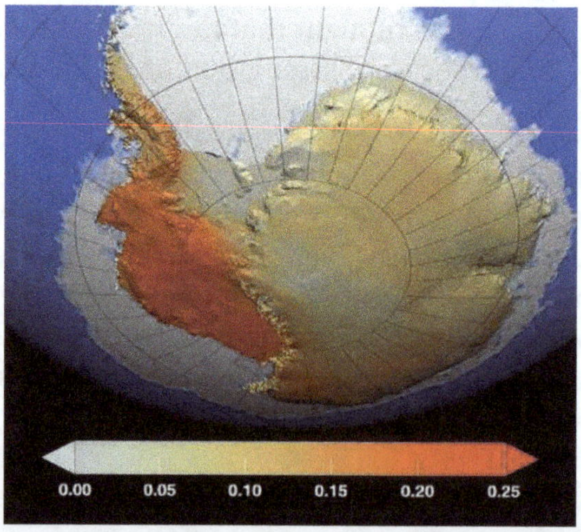

NASA image showing locations on Antarctica where temperatures had increased between 1959 and 2009. Red represents areas where temperatures had increased the most over the period, particularly in West Antarctica, while dark blue represents areas with a lesser degree of warming. Temperature changes are measured in degrees Celsius.

While the current generation of models predicts that such global sea level changes might take several centuries to occur, it is possible that the rate could accelerate as a result of processes that tend to hasten the collapse of ice sheets. One such process is the development of moulins—large vertical shafts in the ice that allow surface meltwater to penetrate to the base of the ice sheet. A second process involves the vast ice shelves off Antarctica that buttress the grounded continental ice sheet of Antarctica's interior. If those ice shelves collapse, the continental ice sheet could become unstable, slide rapidly toward the ocean, and melt, thereby further increasing mean sea level. Thus far, neither process has been incorporated into the theoretical models used to predict sea level rise.

Ocean Circulation Changes

Another possible consequence of global warming is a decrease in the global ocean circulation system known as the "thermohaline circulation" or "great ocean conveyor belt." This system involves the sinking of cold saline waters in the subpolar regions of the oceans, an action that helps to drive warmer surface waters poleward from the subtropics. As a result of this process, a warming influence is carried to Iceland and the coastal regions of Europe that moderates the climate in those regions. Some scientists believe that global warming could shut down this ocean current system by creating an influx of fresh water from melting ice sheets and glaciers into the subpolar North Atlantic Ocean. Since fresh water is less dense than saline water, a significant intrusion of fresh water would lower the density of the surface waters and thus inhibit the sinking motion that drives the large-scale thermohaline circulation. It has also been speculated that, as a consequence of large-scale surface warming, such changes could even trigger colder conditions in regions surrounding the North Atlantic. Experiments with modern climate models suggest that such an event would be unlikely. Instead, a moderate weakening of the thermohaline circulation might occur that would lead to a dampening of surface warming—rather than actual cooling—in the higher latitudes of the North Atlantic Ocean.

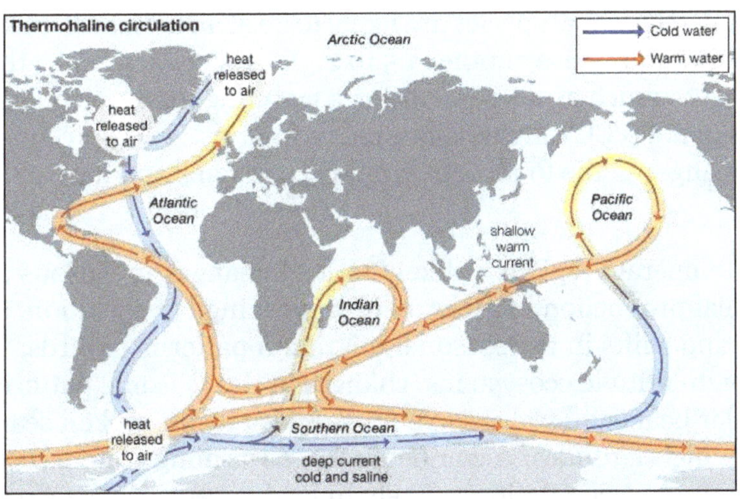

Thermohaline circulation transports and mixes the water of the oceans. In the process it transports heat, which influences regional climate patterns. The density of seawater is determined by the temperature and salinity of a volume of seawater at a particular location. The difference in density between one location and another drives the thermohaline circulation.

Tropical Cyclones

One of the more controversial topics in the science of climate change involves the impact of global warming on tropical cyclone activity. It appears likely that rising tropical ocean temperatures associated with global warming will lead to an increase in the intensity (and the associated destructive potential) of tropical cyclones. In the Atlantic a close relationship has been observed between rising ocean temperatures and a rise in the strength of hurricanes. Trends in the intensities of tropical cyclones in other regions, such as in the tropical Pacific and Indian oceans, are more uncertain due to a paucity of reliable long-term measurements.

While the warming of oceans favours increased tropical cyclone intensities, it is unclear to what extent rising temperatures affect the number of tropical cyclones that occur each year. Other factors, such as wind shear, could play a role. If climate change increases the amount of wind shear—a factor that discourages the formation of tropical cyclones—in regions where such storms tend to form, it might partially mitigate the impact of warmer temperatures. On the other hand, changes in atmospheric winds are themselves uncertain—because of, for example, uncertainties in how climate change will affect ENSO.

Environmental Consequences of Global Warming

Global warming and climate change have the potential to alter biological systems. More specifically, changes to near-surface air temperatures will likely influence ecosystem functioning and thus the biodiversity of plants, animals, and other forms of life. The current geographic ranges of plant and animal species have been established by adaptation to long-term seasonal climate patterns. As global warming alters these patterns on timescales considerably shorter than those that arose in the past from natural climate variability, relatively sudden climatic changes may challenge the natural adaptive capacity of many species.

A large fraction of plant and animal species are likely to be at an increased risk of extinction if global average surface temperatures rise another 1.5 to 2.5 °C (2.7 to 4.5 °F) by the year 2100. Species loss estimates climb to as much as 40 percent for a warming in excess of 4.5 °C (8.1 °F)—a level that could be reached in the IPCC's higher emissions scenarios. A 40 percent extinction rate would likely lead to major changes in the food webs within ecosystems and have a destructive impact on ecosystem function.

Surface warming in temperate regions is likely to lead changes in various seasonal processes—for instance, earlier leaf production by trees, earlier greening of vegetation, altered timing of egg laying and hatching, and shifts in the seasonal migration patterns of birds, fishes, and other migratory animals. In high-latitude ecosystems, changes in the seasonal patterns of sea ice threaten predators such as polar bears and walruses; both species rely on broken sea ice for their hunting activities. Also in the high latitudes, a combination of warming waters, decreased sea ice, and changes in ocean salinity and circulation is likely to lead to reductions or redistributions in populations of algae and plankton. As a result, fish and other organisms that forage upon algae and plankton may be threatened. On land, rising temperatures and changes in precipitation patterns and drought frequencies are likely to alter patterns of disturbance by fires and pests.

Numerous ecologists, conservation biologists, and other scientists studying climate warn that rising surface temperatures will bring about an increased extinction risk. In 2015 one study that examined 130 extinction models developed in previous studies predicted that 5.2 percent of species would be lost with a rise in average temperatures of 2 °C (3.6 °F) above temperature benchmarks from before the onset of the Industrial Revolution. The study also predicted that 16 percent of

Earth's species would be lost if surface warming increased to about 4.3 °C (7.7 °F) above preindustrial temperature benchmarks.

Other likely impacts on the environment include the destruction of many coastal wetlands, salt marshes, and mangrove swamps as a result of rising sea levels and the loss of certain rare and fragile habitats that are often home to specialist species that are unable to thrive in other environments. For example, certain amphibians limited to isolated tropical cloud forests either have become extinct already or are under serious threat of extinction. Cloud forests—tropical forests that depend on persistent condensation of moisture in the air—are disappearing as optimal condensation levels move to higher elevations in response to warming temperatures in the lower atmosphere.

In many cases a combination of stresses caused by climate change as well as human activity represents a considerably greater threat than either climatic stresses or nonclimatic stresses alone. A particularly important example is coral reefs, which contain much of the ocean's biodiversity. Rising ocean temperatures increase the tendency for coral bleaching (a condition where zooxanthellae, or yellow-green algae, living in symbiosis with coral either lose their pigments or abandon the coral polyps altogether), and they also raise the likelihood of greater physical damage by progressively more destructive tropical cyclones. In many areas coral is also under stress from increased ocean acidification, marine pollution, runoff from agricultural fertilizer, and physical damage by boat anchors and dredging.

Another example of how climate and nonclimatic stresses combine is illustrated by the threat to migratory animals. As these animals attempt to relocate to regions with more favourable climate conditions, they are likely to encounter impediments such as highways, walls, artificial waterways, and other man-made structures.

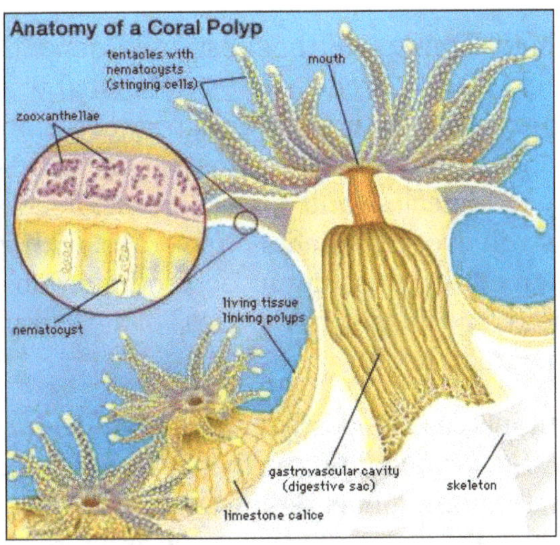

Cross section of a generalized coral polyp.

Warmer temperatures are also likely to affect the spread of infectious diseases, since the geographic ranges of carriers, such as insects and rodents, are often limited by climatic conditions. Warmer winter conditions in New York in 1999, for example, appear to have facilitated an outbreak of West Nile virus, whereas the lack of killing frosts in New Orleans during the early 1990s led to an explosion of disease-carrying mosquitoes and cockroaches. Warmer winters in the Korean peninsula and southern Europe have allowed the spread of the Anopheles mosquito, which carries the malaria parasite, whereas warmer conditions in Scandinavia in recent years have allowed for the northward advance of encephalitis.

In the southwestern United States, alternations between drought and flooding related in part to the ENSO phenomenon have created conditions favourable for the spread of hantaviruses by rodents. The spread of mosquito-borne Rift Valley fever in equatorial East Africa has also been related to wet conditions in the region associated with ENSO. Severe weather conditions conducive to rodents or insects have been implicated in infectious disease outbreaks—for instance, the outbreaks of cholera and leptospirosis that occurred after Hurricane Mitch struck Central America in 1998. Global warming could therefore affect the spread of infectious disease through its influence on ENSO or on severe weather conditions.

Anopheles mosquito, carrier of the malarial parasite.

Socioeconomic Consequences of Global Warming

Socioeconomic impacts of global warming could be substantial, depending on the actual temperature increases over the next century. Models predict that a net global warming of 1 to 3 °C (1.8 to 5.4 °F) beyond the late 20th-century global average would produce economic losses in some regions (particularly the tropics and high latitudes) and economic benefits in others. For warming beyond those levels, benefits would tend to decline and costs increase. For warming in excess of 4 °C (7.2 °F), models predict that costs will exceed benefits on average, with global mean economic losses estimated between 1 and 5 percent of gross domestic product. Substantial disruptions could be expected under those conditions, specifically in the areas of agriculture, food and forest products, water and energy supply, and human health.

Agricultural productivity might increase modestly in temperate regions for some crops in response to a local warming of 1–3 °C (1.8–5.4 °F), but productivity will generally decrease with further warming. For tropical and subtropical regions, models predict decreases in crop productivity for even small increases in local warming. In some cases, adaptations such as altered planting practices are projected to ameliorate losses in productivity for modest amounts of warming. An increased incidence of drought and flood events would likely lead to further decreases in agricultural productivity and to decreases in livestock production, particularly among subsistence farmers in tropical regions. In regions such as the African Sahel, decreases in agricultural productivity have already been observed as a result of shortened growing seasons, which in turn have occurred as a result of warmer and drier climatic conditions. In other regions, changes in agricultural practice, such as planting crops earlier in the growing season, have been undertaken. The warming of oceans is predicted to have an adverse impact on commercial fisheries by changing the distribution and productivity of various fish species, whereas commercial timber productivity may increase globally with modest warming.

Water resources are likely to be affected substantially by global warming. At current rates of warming, a 10–40 percent increase in average surface runoff and water availability has been projected in higher latitudes and in certain wet regions in the tropics by the middle of the 21st century, while decreases of similar magnitude are expected in other parts of the tropics and in the dry regions in the subtropics. This would be particularly severe during the summer season. In many cases water availability is already decreasing or expected to decrease in regions that have been stressed for water resources since the turn of the 21st century. Such regions as the African Sahel, western North America, southern Africa, the Middle East, and western Australia continue to be particularly vulnerable. In these regions drought is projected to increase in both magnitude and extent, which would bring about adverse effects on agriculture and livestock raising. Earlier and increased spring runoff is already being observed in western North America and other temperate regions served by glacial or snow-fed streams and rivers. Fresh water currently stored by mountain glaciers and snow in both the tropics and extratropics is also projected to decline and thus reduce the availability of fresh water for more than 15 percent of the world's population. It is also likely that warming temperatures, through their impact on biological activity in lakes and rivers, may have an adverse impact on water quality, further diminishing access to safe water sources for drinking or farming. For example, warmer waters favour an increased frequency of nuisance algal blooms, which can pose health risks to humans. Risk-management procedures have already been taken by some countries in response to expected changes in water availability.

Energy availability and use could be affected in at least two distinct ways by rising surface temperatures. In general, warmer conditions would favour an increased demand for air-conditioning; however, this would be at least partially offset by decreased demand for winter heating in temperate regions. Energy generation that requires water either directly, as in hydroelectric power, or indirectly, as in steam turbines used in coal-fired power plants or in cooling towers used in nuclear power plants, may become more difficult in regions with reduced water supplies.

It is expected that human health will be further stressed under global warming conditions by potential increases in the spread of infectious diseases. Declines in overall human health might occur with increases in the levels of malnutrition due to disruptions in food production and by increases in the incidence of afflictions. Such afflictions could include diarrhea, cardiorespiratory illness, and allergic reactions in the midlatitudes of the Northern Hemisphere as a result of rising levels of pollen. Rising heat-related mortality, such as that observed in response to the 2003 European heat wave, might occur in many regions, especially in impoverished areas where air-conditioning is not generally available.

The economic infrastructure of most countries is predicted to be severely strained by global warming and climate change. Poor countries and communities with limited adaptive capacities are likely to be disproportionately affected. Projected increases in the incidence of severe weather, heavy flooding, and wildfires associated with reduced summer ground moisture in many regions will threaten homes, dams, transportation networks and other facets of human infrastructure. In high-latitude and mountain regions, melting permafrost is likely to lead to ground instability or rock avalanches, further threatening structures in those regions. Rising sea levels and the increased potential for severe tropical cyclones represent a heightened threat to coastal communities throughout the world. It has been estimated that an additional warming of 1–3 °C (1.8–5.4 °F) beyond the late 20th-century global average would threaten millions more people with the risk of annual flooding. People in the densely populated, poor, low-lying regions of Africa, Asia, and tropical islands would be the most vulnerable, given their limited adaptive capacity. In addition, certain regions in developed countries, such as the Low Countries of Europe and the Eastern Seaboard and Gulf Coast of the United States, would also be vulnerable to the effects of rising sea levels. Adaptive steps are already being taken by some governments to reduce the threat of increased coastal vulnerability through the construction of dams and drainage works.

Permissions

All chapters in this book are published with permission under the Creative Commons Attribution Share Alike License or equivalent. Every chapter published in this book has been scrutinized by our experts. Their significance has been extensively debated. The topics covered herein carry significant information for a comprehensive understanding. They may even be implemented as practical applications or may be referred to as a beginning point for further studies.

We would like to thank the editorial team for lending their expertise to make the book truly unique. They have played a crucial role in the development of this book. Without their invaluable contributions this book wouldn't have been possible. They have made vital efforts to compile up to date information on the varied aspects of this subject to make this book a valuable addition to the collection of many professionals and students.

This book was conceptualized with the vision of imparting up-to-date and integrated information in this field. To ensure the same, a matchless editorial board was set up. Every individual on the board went through rigorous rounds of assessment to prove their worth. After which they invested a large part of their time researching and compiling the most relevant data for our readers.

The editorial board has been involved in producing this book since its inception. They have spent rigorous hours researching and exploring the diverse topics which have resulted in the successful publishing of this book. They have passed on their knowledge of decades through this book. To expedite this challenging task, the publisher supported the team at every step. A small team of assistant editors was also appointed to further simplify the editing procedure and attain best results for the readers.

Apart from the editorial board, the designing team has also invested a significant amount of their time in understanding the subject and creating the most relevant covers. They scrutinized every image to scout for the most suitable representation of the subject and create an appropriate cover for the book.

The publishing team has been an ardent support to the editorial, designing and production team. Their endless efforts to recruit the best for this project, has resulted in the accomplishment of this book. They are a veteran in the field of academics and their pool of knowledge is as vast as their experience in printing. Their expertise and guidance has proved useful at every step. Their uncompromising quality standards have made this book an exceptional effort. Their encouragement from time to time has been an inspiration for everyone.

The publisher and the editorial board hope that this book will prove to be a valuable piece of knowledge for students, practitioners and scholars across the globe.

Index

www.ingramcontent.com/pod-product-compliance
Lightning Source LLC
LaVergne TN
LVHW080655210926
848485LV00003B/382